Synthesis Lectures on Computer Science

The series publishes short books on general computer science topics that will appeal to advanced students, researchers, and practitioners in a variety of areas within computer science.

Wei Xiao · Christos G. Cassandras · Calin Belta

Safe Autonomy with Control Barrier Functions

Theory and Applications

 Springer

Wei Xiao
Computer Science and Artificial Intelligence
Lab
Massachusetts Institute of Technology
Cambridge, MA, USA

Christos G. Cassandras
Division of Systems Engineering
Boston University
Brookline, MA, USA

Calin Belta
Division of Systems Engineering
Boston University
Brookline, MA, USA

ISSN 1932-1228 ISSN 1932-1686 (electronic)
Synthesis Lectures on Computer Science
ISBN 978-3-031-27578-4 ISBN 978-3-031-27576-0 (eBook)
https://doi.org/10.1007/978-3-031-27576-0

This Springer imprint is published by the registered company Springer Nature Switzerland AG
The registered company address is: Gewerbestrasse 11, 6330 Cham, Switzerland

To Ande Xiao, Sujun Zhang, Fei Xiao (W.X.)
To Yu-Chi (Larry) Ho, mentor and friend (C.G.C.)
To Ana and Stefan (C.B.)

Preface

Safety is an essential requirement for any control system, be it a simple home appliance that should not overheat or an aircraft that should not place its passengers' lives at risk. However, *formally guaranteeing* safety (which can informally be stated as "something bad will never happen") has not been a top requirement for designing controllers as long as testing for safe operation through prototypes or simulation experiments is practically feasible. The rapid proliferation of large-scale complex cyber-physical systems over the past two decades has made such testing infeasible and the emergence of *autonomous* systems has further made *safety* a critical design requirement. The "internet of vehicles," envisioned as the next generation of intelligent transportation systems where self-driving cars cooperate to eliminate accidents (94% of which are due to human error) and to reduce congestion and harmful emissions, is a prime example of such a large-scale system consisting of interacting autonomous entities (here, the vehicles, are generically referred to as "agents"). In fact, in today's technologically insatiable society, autonomous systems are fast becoming ubiquitous, from autonomous driving to sophisticated manufacturing systems and automated delivery of goods or exploration of hazardous environments.

The goal of this book is to contribute to safe autonomy becoming a critical component of control system design. A central theme in its content is enhancing our understanding of the interplay between safety and the other two critical control design components: *performance* and *computational complexity*. Indeed, one can always design an overly conservative safe system, such as a fleet of autonomous vehicles moving at no more than 10 m/h, leading to unacceptably poor system performance. On the other hand, pushing the limits of safety with a sophisticated controller, such as autonomous vehicles cruising at 100 m/h while ensuring safe operation, would require a prohibitively high computational cost preventing real-time control operation or an intolerably high economic cost in terms of the equipment required to guarantee safety.

In this book, we present a framework based on *Control Barrier Functions* (CBFs) that endows a control system with the ability to guarantee the satisfaction of safety requirements as formally specified by the designer in the form of inequalities involving state variables defined in the model of the dynamic system being controlled. A key attractive feature of this framework is the ability to transform these inequalities into a new set of conditions which allow the specification of a set of feedback control laws whose implementation implies the satisfaction of the original safety requirements. Moreover, determining these controls is computationally tractable and often quite simple. Ultimately, we combine the well-established theory of constrained optimal control with CBFs to achieve near-optimal system performance while also guaranteeing the safety requirements at all times thanks to a forward invariance property intrinsic in the CBF approach.

Naturally, there is no free lunch in this framework. The formal safety guarantees come at the price of some conservativeness, setting the stage for the trade-offs discussed earlier involving safety, performance, and computational complexity. Moreover, the implementation of a CBF-based controller in real time faces a number of challenges, central among them being the ability to always determine a *feasible* control action at any time when the controller is designed to operate in a discretized time setting. The book explores various techniques, including the use of novel *event-driven* methods, to confront this challenge and largely alleviate it. It also embraces *machine learning* methods which can further contribute to addressing the feasible control implementation problem.

The promise of autonomous systems, in terms of their capabilities and societal contributions for this century, is extremely exciting. Their complexity, on the other hand, can be overwhelming and their deployment is hindered by the need to provide them with guarantees for a safe operation since they directly involve humans, as in transporting passengers through autonomous vehicles or industrial equipment working cooperatively with human operators. Powerful new formal methodologies are needed to ensure safe autonomy and to deliver the full potential of these systems.

About This Book

The majority of this book is based on a series of recent papers published by the authors in leading journals and conference proceedings within the systems and control community. A substantial part also comes from the 2021 doctoral dissertation of the first author entitled "Optimal Control and Learning for Safety-Critical Autonomous Systems."

The book is intended for an audience of graduate-level students with a mature understanding of basic control theory and optimization methods, as well as for researchers who are concentrating on problems involving autonomous systems and their applications to areas such as robotics and autonomous vehicles in transportation systems. We also envision it as a textbook for a graduate course that can provide a first exposure to concepts of both autonomy and safety, as well as a detailed exposition of the theory of Control Barrier

Functions (CBFs). Throughout the book, we include examples of how this theory can be used, including extensive simulations of challenging autonomous systems and comparisons with "baseline" controllers that can illustrate the benefits of the specific methods we develop.

We have made every effort to make the book as self-contained as possible, with the understanding that the reader has a background on basic dynamic system modeling, control theory, and essential elements of optimization, including fundamentals of optimal control theory.

Book Organization

The basic road map and organization approach for the book are briefly described next.

Chapter 1 provides a brief introduction to the concepts of *autonomy* and *safety*. It sets the stage for one of the ultimate goals of the book: controlling autonomous systems so as to combine safety with application-dependent control objectives. It also places the design objective of "safety" in the context of inevitable trade-offs with the other two objectives: performance and computational complexity. Readers familiar with basic optimal control theory will recognize part of this chapter as reviewing material to establish key problem formulations and notation to be used throughout the remaining chapters.

The next nine chapters of the book are logically organized in three parts. First, Chaps. 2 and 3 lay out the basic background material involving barrier functions leading to the definitions of Control Barrier Functions (CBFs) and High-Order Control Barrier Functions (HOCBFs). Next, Chaps. 4–8 address the challenges that the CBF/HOCBF-based control design must address in order for controllers to be fully implementable and effectively combined with optimal control theory. Finally, Chaps. 9 and 10 present a variety of applications of the CBF/HOCBF-based method for designing safe controllers for autonomous systems encountered in traffic networks and in robotics.

Chapter 2 introduces *Control Barrier Functions* (CBFs) in the context of control systems, tracing back the history of CBFs to the barrier functions used in classical mathematical optimization. These can be extended to Lyapunov-like barrier functions and ultimately adapted to the control of dynamical systems in the presence of constraints. Formal definitions of all the concepts used in the rest of the book are provided in detail. A first explanation of how CBFs can be combined with constrained optimal control problems introduced in Chap. 1 is also presented, along with the formal general-purpose solution methodology that will be used for such problems through-out the book. In particular, it is shown how these problems can be converted into a sequence of Quadratic Programs (QPs) once the time line is appropriately discretized. Chapter 3 is motivated by a potential limitation in the definition of a CBF: the function capturing the safety requirement we are interested in may have a relative degree greater than one (i.e., its first derivative may be independent of the control variables). This limitation is relaxed by a more general definition which leads to *High-Order Control Barrier Functions* (HOCBFs). These are even further generalized to include *integral HOCBFs* (iHOCBFs).

It is recommended that Chaps. 2 and 3 form the content of a course that introduces the theory of safe autonomy using CBFs. By the end of these two chapters, the reader will have grasped the basic concepts and methodologies involved in designing CBFs (or HOCBFs, as needed) and solving constrained optimal control problems where the constraints are in the specific form of CBF/HOCBF-based inequalities. The reader will also have gained an appreciation of the power of this methodology (illustrated through several examples), but also the technical challenges it entails, specifically the potential conservativeness that CBFs may bring to control design and the issue of always finding a feasible control action when implementing CBF-based controllers. The second part of the book, consisting of Chaps. 4–8, addresses these two main challenges.

Chapter 4 addresses the crucial issue of whether feasible solutions can be found when deriving an explicit solution of an optimal control problem through a sequence of QPs as explained in Chap. 2. The problem of feasibility stems from the potential conflict between a CBF/HOCBF-based constraint and control limits that are unavoidable in practice. It also comes from the inter-sampling effect due to the time discretization, as there are rarely reliable guidelines for the choice of the time step used in any such process. Two methods are presented, one based on assigning adjustable penalties to the CBF/HOBF-based constraints and the other on designing a special additional CBF, which leads to a sufficient condition for feasibility.

Chapter 5 is a continuation of Chap. 4's effort to address the feasibility challenge in implementing CBF/HOCBF-based controllers. Machine learning methodologies are utilized to take advantage of actual data that are generally available from the operation of an autonomous system. A "feasibility robustness" metric is introduced to measure the extent to which the feasibility problem we are concerned with is maintained in the presence of time-varying and unknown unsafe sets. The main idea rests on parameterizing the CBFs/HOCBFs involved in a safe-critical control problem and then learning the parameter values that maximize this feasibility robustness metric. The reader is expected to possess a basic knowledge of *machine learning* algorithms for classification problems.

Chapter 6 focuses on the issue of conservativeness which may be introduced when a CBF/HOBF-based controller is designed. To deal with this issue, we introduce *adaptive CBFs* (aCBFs) to guarantee safety and feasibility under time-varying control bounds, as well as potentially noisy dynamics. Two different forms of aCBFs are described: *parameter-adaptive* CBFs and *relaxation-adaptive* CBFs. Although there is a remote connection of aCBFs to traditional adaptive control methods, the reader is not expected to have any knowledge of such methods since this chapter considers only parameter uncertainties rather than unknown system dynamics, a topic studied in the next chapter. Chapter 7 considers systems with unknown dynamics, including the case of noise present in the system. Here, we introduce an *event-driven* (or *event-triggered*) framework that also deals with the same feasibility problem studied in Chaps. 4 and 5. It is shown that this approach can almost entirely eliminate the feasibility issues of concern by selecting specific events to trigger the execution of a QP that leads to a control action update, hence eliminating the need for specifying a time step value needed in the time-driven controller implementation.

Chapter 8 presents a two-step procedure that combines traditional optimal control and CBF/HOCBF methods to achieve both optimality and guaranteed safety in real-time control. This can be viewed as a way to bridge the gap between planning a trajectory and actually executing it in the presence of noise and model uncertainties. The key idea is to first generate trajectories by solving a tractable optimal control problem and then track these trajectories using a controller that simultaneously ensures that all constraints are satisfied at all times. This is termed an *Optimal control with Control Barrier Functions* (OCBF) controller. This chapter may be read after Chaps. 2 and 3 and does not require a deep understanding of Chaps. 4–7. It does, however, require a basic knowledge of constrained optimal control theory.

The last part of the book, consisting of Chaps. 9 and 10, is intended to provide the reader with specific applications of the CBF/HOCBF methodology developed throughout the earlier chapters. These last two chapters heavily rely on the OCBF approach presented in Chap. 8.

Chapter 9 considers safety-critical optimal control problems encountered in the next-generation intelligent transportation systems that have started to emerge. In particular, these problems involve the cooperative coordination of Connected Autonomous Vehicles (CAVs) in traffic networks. Three specific applications are presented in detail: traffic merging, traffic control in roundabouts, and traffic control in multi-lane signal-free intersections.

Chapter 10 looks at three applications of the use of the CBF/HOCBF methodology encountered in robotics: a ground-robot executing typical two-dimensional navigation tasks where obstacle avoidance is crucial, autonomous driving where a vehicle must satisfy complex traffic law specifications, and safe navigation tasks for an aerial vehicle (quadrotor).

Cambridge, MA, USA
Boston, MA, USA
January 2023

Wei Xiao
Christos G. Cassandras
Calin Belta

Acknowledgements

The authors would like to collectively acknowledge a number of colleagues and students who have substantially contributed to the state of the art in safety-critical control and the theory of CBFs in particular. Their work has helped us build an understanding of the foundations based on which our own work has evolved and has led to this book. We acknowledge the fundamental work by Mitio Nagumo back in 1942, as well as more recent work on barrier certificates by Ali Jadbabaie, George Pappas, and Miroslav Krstic. We became familiar with the original CBF concept, which is the starting point in this book, through the work of Aaron Ames, Dimos Dimarogonas, Magnus Egerstedt, Jessy Grizzle, Paulo Tabuada, and Koushil Sreenath, and their colleagues. In addition, we would like to thank Daniela Rus for extensive discussions on safety-critical systems in many areas. Last but not least, we would like to thank our collaborators in projects related to CBF-based research: Makram Chahine, Rui Chen, Max Cohen, Emilio Frazzoli, Ramin Hasani, Andreas Malikopoulos, Noushin Mehdipour, Nader Meskin, Yannis Paschalidis, Ehsan Sabouni, Radboud Duintjer Tebbens, Tsun-Hsuan Wang, Huile Xu, Kaiyuan Xu, and Yue Zhang. The research areas investigating CBFs and their applications are growing fast and there is undoubtedly related work that has not been covered in this book.

Finally, the authors would like to express their gratitude to the following sponsors of their research which has contributed to this book: the National Science Foundation, the Air Force Office of Scientific Research, ARPA-E, Motional, and The MathWorks.

Cambridge, MA, USA
Boston, MA, USA
January 2023

Wei Xiao
Christos G. Cassandras
Calin Belta

Contents

About the Authors

Wei Xiao Ph.D. is a Postdoctoral Associate at the Massachusetts Institute of Technology. He received a B.Sc. from the University of Science and Technology Beijing, a M.Sc. degree from the Chinese Academy of Sciences (Institute of Automation), and a Ph.D. from Boston University. His research interests include control theory and machine learning with particular emphasis on robotics and traffic control. He received an Outstanding Student Paper Award at the 2020 *IEEE Conference on Decision and Control.*

Christos G. Cassandras Ph.D. is a Distinguished Professor of Engineering at Boston University. He is Head of the Division of Systems Engineering, Professor of Electrical and Computer Engineering, and Co-Founder of Boston University's Center for Information and Systems Engineering (CISE). He received a B.S. from Yale University, a M.S.E.E from Stanford University, and S.M. and Ph.D. degrees from Harvard University. He specializes in the areas of discrete event and hybrid systems, cooperative control, stochastic optimization, and computer simulation, with applications to computer and sensor networks, manufacturing systems, and transportation systems. He has published over 500 refereed papers in these areas and 7 books. He has guest-edited several technical journal issues and serves on several journal editorial boards. In addition to his academic activities, he has worked extensively with industrial organizations on various systems integration projects and the development of decision-support software. He has most recently collaborated with MathWorks, Inc. in the development of the discrete event and hybrid system simulator SimEvents. He is a Fellow of the IEEE and a Fellow of the IFAC.

Calin Belta Ph.D. is a Professor in the Department of Mechanical Engineering at Boston University, where he holds the Tegan family Distinguished Faculty Fellowship. He is also the Director of the BU Robotics Lab. He received B.Sc. and M.Sc. degrees from the Technical University of Iasi and M.Sc. and Ph.D. degrees from the University of Pennsylvania.

His research interests include dynamics and control theory, with particular emphasis on hybrid and cyber-physical systems, formal synthesis and verification, and applications in robotics and systems biology. He has received the Air Force Office of Scientific Research Young Investigator Award and the National Science Foundation CAREER Award. He is a Fellow and Distinguished Lecturer of IEEE.

Introduction to Autonomy and Safety

1

Autonomous systems have emerged as a consequence of multiple technological developments including (i) ubiquitous wireless networking capabilities allowing a multitude of *agents* (ranging from miniature sensing devices to large vehicles) to share information reliably and efficiently, (ii) sensors with different possible modalities that can be placed on board the agents giving them unprecedented environment perception capabilities, and (iii) smaller and faster processors capable of on-board computational tasks that were unimaginable a couple of decades ago.

What exactly do we mean when we refer to a system as *autonomous*? The answer largely depends on the application domain, but there are some characteristics of autonomy that one can identify in terms of the goals that the system aims to achieve, the information that its control actions are based on, and its relationship to the humans who have designed it and expect a certain type and level of operational performance. Thus, for starters, an autonomous system is designed to *achieve a given set of goals which may vary over time*, partly based on how the environment within which the system operates changes. To do so, the components of such a system (which we will generically refer to as *agents*) must be able to sense and collect information from the environment and possibly share it. It follows that an important feature of an autonomous system is that each agent has the ability to both operate separately from others but also *cooperate* with them in terms of information sharing, goal setting and control action execution. Finally, it is expected that an autonomous system operate over long periods of time unattended, i.e., without human intervention. An example of an autonomous system is the so-called "internet of vehicles" consisting of self-driving cars which can deliver passengers (or goods) to specific destinations by controlling their routes as well as their powertrains in a way that does not conflict (and in fact cooperates) with other vehicles, as illustrated in Fig. 1.1. Other examples include teams of mobile robots that cooperate to carry out various tasks such as performing fundamental storage-retrieval

© The Author(s), under exclusive license to Springer Nature Switzerland AG 2023
W. Xiao et al., *Safe Autonomy with Control Barrier Functions*,
Synthesis Lectures on Computer Science,
https://doi.org/10.1007/978-3-031-27576-0_1

operations in a warehouse, monitoring an environment to detect different types of events such as fires in wooded areas, seeking for faults in a bridge, etc.

Fig. 1.1 Autonomy in the form of vehicles employing connectivity, sensing, and automatic control

While an autonomous system is an exciting prospect for 21st century technology, ensuring that it always operates as "intended" poses significant challenges. At a minimum, one expects that such a system is always guaranteed a minimum level of performance that includes some quantifiable metrics of safety. For example, if we are to trust a self-driving vehicle, we need to be absolutely certain that it does not violate the basic driving rules that a reasonable human driver would always obey. Beyond that, we also expect the vehicle to be able to outperform a human driver in terms of accident prevention, such as detecting and avoiding obstacles faster and more effectively than any human driver. Given the sensing capabilities, high-speed processing of information, and intelligent decision making potential of an autonomous system agent, we should expect that it can easily outperform its human counterpart operator. However, we first need to develop a rigorous framework that can provide such *safety guarantees*.

The traditional setting for studying dynamic systems subject to constraints while aiming to achieve the best possible performance is provided by optimal control theory. This has been extensively developed for both discrete and continuous time models. In a continuous time setting, the performance of the system is measured through a scalar function

$$J = \int_{t_0}^{t_f} L[x(t), u(t), t]dt + \phi[x(t_f), t_f], \tag{1.1}$$

where $x(t)$ is the state of the system, which evolves over the time interval $[t_0, t_f]$ according to

$$\dot{x} = f[x(t), u(t), t] \tag{1.2}$$

with a given initial condition $x(t_0)$. In the above equation, $\dot{()}$ denotes differentiation with respect to time and f is a function describing the (possibly time-varying) dynamics of the system. The problem is to determine functions $u(t)$ (i.e., the *controls*) that minimize (or maximize) J. Usually, the state and control are subject to various constraints of the general form

$$g[x(t), u(t), t] \leq 0 \quad \forall t \in [t_0, t_f] \tag{1.3}$$

or possibly

$$g[x(t_k), u(t_k), t_k] \leq 0 \quad t_k \in [t_0, t_f], \tag{1.4}$$

where t_k can be an isolated time instant or can belong to subintervals of $[t_0, t_f]$. The constraint in (1.4) includes the case of specified terminal states, i.e., $x(t_f) - x_f = 0$. It is worth noting that in many cases these constraints do not explicitly include any control component. For example, maintaining a minimum distance from the side of the road in autonomous driving can be written as an inequality over the vehicle's position, which is part of the vehicle's state, and does not explicitly involve controls, such as acceleration.

The constraints in (1.3) or (1.4) can be *hard* or *soft*. Hard constraints (e.g., "avoid collisions with other traffic participants" in autonomous driving) are usually referred to as *safety constraints*, with the understanding that the solution $u(t)$ to the problem needs to strictly enforce them. Soft constraints (e.g., "stay in lane" in autonomous driving) can be slightly violated, even though this should be avoided if possible. This can be handled in a number of different ways. If, for example, there are state constraints to be satisfied only at the terminal time t_f, then they can be incorporated in $\phi[x(t_f), t_f]$ in (1.1). Alternatively, they can be expressed in the "softer form":

$$g[x(t), u(t), t] \leq \delta \quad \forall t \in [t_0, t_f], \tag{1.5}$$

where $\delta \geq 0$ is an additional parameter to be minimized alongside the original cost.

In this book, we deal with both hard and soft constraints, but we often focus on the hard *safety constraints* that must always be satisfied if we are to design *safe* autonomous systems. For multi-agent autonomous systems, the state $x(t)$ is usually expressed as a vector $[x_1(t), \ldots, x_N(t)]$ where $x_i(t)$, $i = 1, \ldots, N$ is the state (generally, also a vector) that corresponds to the ith agent in the system. In other words, the system possesses a structure consisting of multiple interacting agents. Consequently, many of the constraints apply to individual agents or to subsets of agents. Along the same lines, the function $L[x(t), u(t), t]$ in (1.1) is usually expressed as a weighted sum of performance metrics for individual agents; alternatively, one needs to solve multiple optimal control problems with costs of the form (1.1).

The optimal control framework provides a complete and elegant treatment of problems formulated so as to achieve the best possible system performance while satisfying constraints, which can be expressed in the form (1.3) or (1.4). Unfortunately, obtaining explicit solutions to such problems is the exception rather than the rule. Even when this is possible, it may require a prohibitively expensive computational effort for any applications of these solution techniques to be used on line. In the case of autonomous systems that consist of multiple interconnected networks, the dimensionality of the problems rapidly increases with the inclusion of more agents and one confronts the common non-scalability challenge. In fact, since optimal control problems can always be solved (in principle) through dynamic programming techniques, one quickly has to confront the familiar "curse of dimensionality" coined by Bellman.

It is for these reasons that optimal control theory [12] alone cannot provide us with the means for practically addressing problems of this type. One can of course forego the goal of optimality and compromise by simply seeking feasible solutions that only satisfy (1.3) or (1.4) subject to the system dynamics (1.2). Alternatively, one can seek to derive approximate solutions to the optimization problems while still ensuring that all constraints are satisfied [5]. A popular approach is Model Predictive Control (MPC) [51], in which the optimization is iteratively performed over short receding horizons. MPC has also been shown to work in the presence of uncertainties, while ensuring constraint satisfaction [28, 32, 35, 60].

Safety, which can informally be stated as "something bad will never happen," can be formally expressed as a formula in a rich temporal logic, such as Linear Temporal Logic (LTL) [6, 16] or Signal Temporal Logic (STL) [37]. Such logics also allow us to express specifications such as liveness "something good will eventually happen," and, in principle, any natural language statement that can be translated into Boolean and temporal statements over features of interest [8, 40, 59]. In principle, at the expense of significantly increasing the complexity of the optimization problem, such rich specifications can be mapped onto safety-like requirements [7]. For this reason, and to keep the exposition simple, here we only focus on safety.

In this book, we present a framework that guarantees the satisfaction of a set of safety requirements by an autonomous system, while at the same time involving a reasonable computational effort for online implementations. Central to our approach are *Control Barrier Functions* (CBFs) as outlined in Fig. 1.2. We combine optimal control techniques with the theory of CBFs. We will see, for example, that we can seek simple and efficient solutions

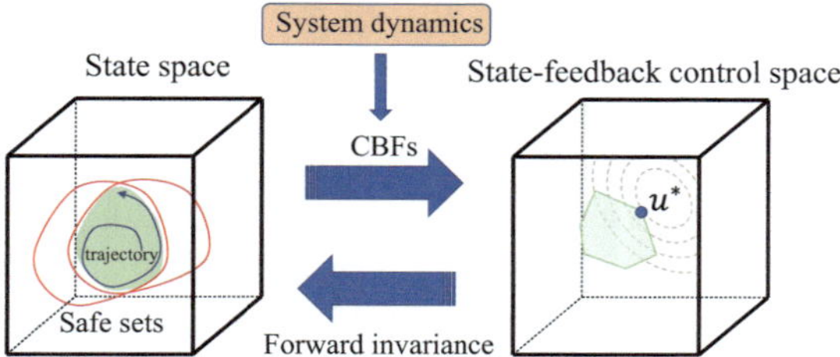

Fig. 1.2 Schematic representation of how CBFs allow us to perform real-time convex optimization, as they map safety constraints from the state space to the control space combining system dynamics. The system trajectory will always stay within the safe green area (a subset of the intersection of safe sets) in the state space due to the forward invariance property of the CBF method. Control subject to CBFs is guaranteed to be safe at the expense of potential suboptimality and conservativeness

Fig. 1.3 The focus of this book with respect to classical control problems

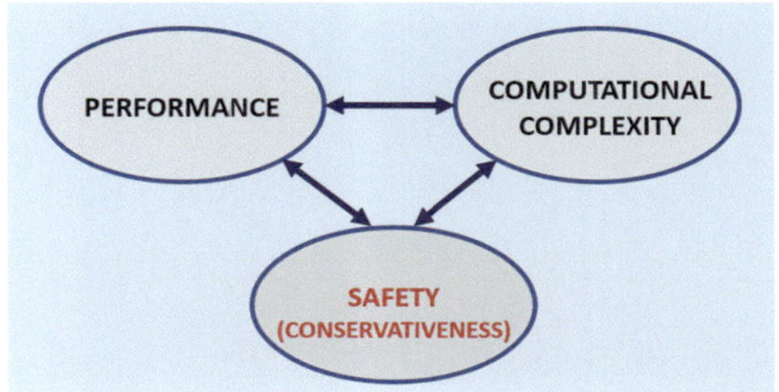

of optimal control problems that ignore safety constraints, just to establish lower (or upper) bounds to what is achievable by the system as it stands. We can then combine these (generally infeasible) solutions with CBF-based methods to get as close as possible to these unconstrained solutions, while at the same time guaranteeing the satisfaction of all safety constraints. Naturally, this approach is valuable only if it can be implemented in a real-time context and with relatively limited inter-agent communication.

Finally, Fig. 1.3 places the focus of this book in perspective relative to classical control problems in which we seek to design control systems that achieve a desired behavior without excessive computational complexity that would prohibit their use in real-time applications. Clearly, there is a trade-off between these two objectives which we must constantly take into account. However, in autonomous systems, safety becomes such a crucial objective that we are often willing to sacrifice performance for the sake of safety, thus sometimes designing the system in an over-conservative manner. Therefore, in addition to trading off performance with computational complexity, the inclusion of hard constraint (safety) guarantees creates a new dimension in control design: the level of conservativeness in the control which also depends on the computational complexity we are willing to tolerate to achieve better performance and less conservativeness.

Control Barrier Functions

2

2.1 Barrier Functions

2.1.1 Barrier Functions in Optimization

In the field of mathematics known as constrained optimization [44] [9], a barrier function is a continuous function whose value increases to infinity at the boundary of the feasible region (i.e., the set of states that satisfy the constraints) in a given optimization problem. Such functions are used to replace inequality constraints by a term in the objective function that incurs a penalty that is easier to handle, typically by requiring only standard calculus-based methods. To see how this is accomplished, consider the following scalar constrained optimization problem:

$$\underset{x \in \mathbb{R}}{\text{minimize}}\ h(x)$$
$$\text{subject to:}\ x \le a \tag{2.1}$$

where $a \in \mathbb{R}$ is a constant, $x \in \mathbb{R}$ is a decision variable, and $h : \mathbb{R} \to \mathbb{R}$ is continuously differentiable. If we wish to remove the above inequality constraint, with the help of an additional function $d : \mathbb{R} \to \mathbb{R}$, we can reformulate the problem as:

$$\underset{x \in \mathbb{R}}{\text{minimize}}\ [h(x) + d(x)],$$
$$\text{where}\ d(x) = 0\ \text{if}\ x \le a,\ \text{and}\ d(x) = \infty\ \text{otherwise.} \tag{2.2}$$

The above problem is equivalent to (2.1), while eliminating the inequality constraint. However, the objective function becomes discontinuous (d is not continuous at a), which prevents us from using standard calculus techniques to solve it.

© The Author(s), under exclusive license to Springer Nature Switzerland AG 2023
W. Xiao et al., *Safe Autonomy with Control Barrier Functions*,
Synthesis Lectures on Computer Science,
https://doi.org/10.1007/978-3-031-27576-0_2

In order to bypass the discontinuity issue, we may approximate the function $d(x)$ in (2.2) by using a properly chosen, continuously differentiable function $c(x)$. Then, we can reformulate the optimization problem (2.2) as:

$$\underset{x \in \mathbb{R}}{\text{minimize}} \; [h(x) + \mu c(x)], \tag{2.3}$$

where $\mu > 0$ is a free parameter. The function c is called a *barrier function*. There are several choices for $c(x)$ in the literature. Here, we limit ourselves to two such functions. First, a logarithmic function $c(x) = -log(a - x)$ can be used. In this case, $c(x)$ is called a "logarithmic barrier function". Note that $c(x) \to \infty$ as $a - x \to 0$ from the positive side. An alternative common choice is $c(x) = \frac{1}{a-x}$. Note that the resulting optimization problems are not equivalent to the original one (2.1), but, as $\mu \to 0$, they provide increasingly better approximations. Such constraints that are included in the cost function are sometimes referred to as "soft" constraints, and are also used in optimization and learning for dynamical systems [10].

2.1.2 Barrier Functions for Safety Verification

We now consider the case where $x \in \mathbb{R}^n$ is the state of a dynamical system, i.e., its time evolution satisfies the following differential equation:

$$\dot{x} = f(x), \tag{2.4}$$

where $f : \mathbb{R}^n \to \mathbb{R}^n$ is assumed to be locally Lipschitz continuous in order to ensure the existence and uniqueness of a solution. Moreover, as is often the case, we wish to ensure that the state of (2.4) always satisfies a *safety constraint* (specification):

$$b(x) \geq 0, \tag{2.5}$$

where $b : \mathbb{R}^n \to \mathbb{R}$ is continuously differentiable.

At this point, we introduce a fundamental result known as Nagumo's theorem [42]. This theorem states that if system (2.4) is initially safe, i.e., $b(x(0)) \geq 0$, then it is always safe, i.e., $b(x(t)) \geq 0, \forall t \geq 0$, if and only if

$$\dot{b}(x) \geq 0, \; \text{whenever } b(x) = 0. \tag{2.6}$$

If $b(x)$ satisfies this condition, then we call $b(x)$ a *barrier function* for system (2.4). In the special case where $\dot{b}(x) = 0$ when $b(x) = 0$, the system state x evolves on the boundary of what we shall henceforth call the *safe set* defined by $\{x \in \mathbb{R} : b(x) \geq 0\}$.

The safety condition (2.6) in Nagumo's theorem is widely used in safety verification and referred to as a *certificate* [48] [70] for dynamical system (2.4). It is also used in a stochastic setting [49], where the system state is treated as a random process. This verification method generally yields what is usually referred to as a *barrier certificate*.

Such barrier certificates can also be used to verify the safety of control systems of the form:

$$\dot{x} = f(x, u), \tag{2.7}$$

where $f : \mathbb{R}^n \times \mathbb{R}^q \to \mathbb{R}^n$ is locally Lipschitz, and $u \in \mathbb{R}^q$ is the control input. Given a state feedback controller $u = k(x)$, where $k : \mathbb{R}^n \to \mathbb{R}^q$, observe that the control system (2.7) can be written in closed form as $\dot{x} = f(x, k(x))$, which is similar to (2.4). Then, we can use the same safety condition (2.6) in Nagumo's theorem to check whether the control system (2.7) under the controller $u = k(x)$ is safe. However, one obvious problem with barrier certificates is that we still do not know how to explicitly find such safe controllers. This can be addressed by using a Lyapunov-like barrier function and subsequently a Control Barrier Function (CBF). These are introduced in the following sections, along with the foundations of what constitutes the *CBF method*.

2.2   Lyapunov-Like Barrier Functions

The safety condition (2.6) in Nagumo's theorem is only defined at the boundary of the safe set. This means that the condition might be discontinuous, analogous to what we saw with the barrier function in the optimization problem (2.2). This discontinuity could prevent us from finding a safe controller.

Example 2.1 Consider an *Adaptive Cruise Control* (ACC) problem, where an autonomous vehicle, called *ego*, must maintain a safe distance from its preceding vehicle. The ego's dynamics are defined as

$$\dot{z} = v_p - v, \quad \dot{v} = u, \tag{2.8}$$

where $z \in \mathbb{R}$ denotes the along-lane distance between the ego vehicle along-lane location $x \in \mathbb{R}$ and its preceding vehicle along-lane location $x_p \in \mathbb{R}$ such that $z = x_p - x$, and $v > 0$, $v_p > 0$ denote the velocities of the ego and its preceding vehicle (the latter assumed to move with constant speed), respectively; $u \in \mathbb{R}$ denotes the control (acceleration) of the ego vehicle.

The safety constraint between the ego and its preceding vehicle requires the distance z to be no less than $l_0 > 0$, i.e., $z \geq l_0$. We then define $b(x) = z - l_0$, where $x = (z, v)$. Nagumo's theorem requires that $\dot{b}(x) = \dot{z} = v_p - v \geq 0$ when $b(x) = 0$ (i.e., $z = x_p - x = l_0$) in order to guarantee safety. In other words, the speed of the ego vehicle should not be higher than its preceding vehicle when $z = l_0$. This is obviously impossible in most cases, as we cannot instantaneously change the speed of the ego vehicle when the distance between it and its preceding vehicle is l_0. This sudden speed change requirement corresponds to the discontinuity in finding a safe controller for ACC. This discontinuity is equivalent to a change in speed within an infinitely small time interval, and, therefore, requires infinitely large deceleration, which is practically impossible.   ∎

In order to address this discontinuity problem, we may strengthen the safety condition (2.6) in Nagumo's theorem, in a manner similar to what we did for the optimization problem (2.3). This is fully described in [2] [21]. In short, we modify (2.6) into the following strengthened safety condition:

$$\dot{b}(x) + \epsilon b(x) \geq 0, \tag{2.9}$$

where $\epsilon > 0$ is a constant. It is clear from the above equation that whenever $b(x) = 0$, we recover the original condition $\dot{b}(x) \geq 0$. Note that (2.9) allows $b(x)$ to decrease inside the safe set so that $\dot{b}(x) \geq -\epsilon b(x)$ when $b(x) > 0$. This means that $\dot{b}(x)$ could be negative inside the safe set. Returning to the above ACC example, the strengthened safety condition is $v_p - v + \epsilon(z - l_0) \geq 0$, which is a continuous speed requirement for the ego vehicle, hence tractable in terms of finding a safe controller; we will return to this example and make this point explicit in subsequent chapters.

Clearly, the safety condition (2.9) is not equivalent to the safety condition (2.6) in Nagumo's theorem, and it is only a sufficient condition for safety, i.e., $b(x(t)) \geq 0, \forall t \geq 0$. This means that this strengthening introduces a measure of *conservativeness*, which could limit the system performance (such as optimality). One of the goals of this book is to address this conservativeness problem. Nevertheless, the strengthened safety condition (2.9) is extremely useful in helping us find a safe controller. This will be further discussed in the next section.

Another way to look at the safety condition (2.9) is from a Lyapunov perspective by letting $V(x) = -b(x)$. Then, the safety condition (2.9) can be rewritten as:

$$\dot{V}(x) + \epsilon V(x) \leq 0. \tag{2.10}$$

The above constraint is exactly the Lyapunov condition for stability [26] [1]. This is the reason why we call a function $b(x)$ that satisfies (2.9) a *Lyapunov-like barrier function*.

For the Lyapunov constraint (2.10), suppose that $b(x(0)) < 0$, i.e., the system state is initially outside the safe set. Therefore, $V(x(0)) = -b(x(0)) > 0$. Following standard Lyapunov stability theory, this means that the system state will be stabilized to the safe set (specifically, the safe set boundary $b(x) = 0$) if the system is initially unsafe. In this case, a Lyapunov-like barrier function becomes a Lyapunov function if we shrink the safe set to a point. Further, if we choose $b(x)$ in (2.9) as a power function with power less than 1, we can find some finite time convergence Lyapunov-like barrier functions [57] [76]. In other words, if (2.9) is satisfied, the system state will be stabilized to the safe set within a specific time bound determined by how we choose the power function.

2.3   Control Barrier Functions

In this section, we formally introduce a Control Barrier Function (CBF) by considering Lyapunov-like barrier functions in a general setting for control systems. We begin with the following definition:

Definition 2.1 (*Class $\mathcal{K}$ and extended class $\mathcal{K}$ functions* [26]) A Lipschitz continuous function $\alpha : [0, a) \to [0, \infty), a > 0$ is said to belong to class $\mathcal{K}$ if it is strictly increasing and $\alpha(0) = 0$. Moreover, $\alpha : [-b, a) \to [-\infty, \infty), a > 0, b > 0$ belongs to extended class $\mathcal{K}$ if it is strictly increasing and $\alpha(0) = 0$.

We first consider affine control systems (non-affine control systems will be considered in the next chapter) of the general form:

$$\dot{x} = f(x) + g(x)u, \tag{2.11}$$

with $x \in X \in \mathbb{R}^n$ (X denotes a closed state constraint set), $f : \mathbb{R}^n \to \mathbb{R}^n$ and $g : \mathbb{R}^n \to \mathbb{R}^{n \times q}$ are locally Lipschitz, and $u \in U \subset \mathbb{R}^q$ (U denotes a closed control constraint set). Solutions $x(t), t \geq 0$, of (2.11) starting at $x(0)$ are forward complete.

Definition 2.2 (*Forward invariant* set) A set $C \subset \mathbb{R}^n$ is forward invariant for system (2.11) if its solutions starting at any $x(0) \in C$ satisfy $x(t) \in C$ for $\forall t \geq 0$.

Throughout this book, we consider a set C (the safe set) defined using a continuously differentiable function $b : \mathbb{R}^n \to \mathbb{R}$ as follows:

$$C := \{x \in \mathbb{R}^n : b(x) \geq 0\}, \tag{2.12}$$

Definition 2.3 (*Control barrier function* [2], [21]) Given a set C as in (2.12), $b(x)$ is a Control Barrier Function (CBF) for system (2.11) if there exists a class $\mathcal{K}$ function α such that

$$\sup_{u \in U}[L_f b(x) + L_g b(x)u + \alpha(b(x))] \geq 0, \quad \forall x \in C, \tag{2.13}$$

where L_f, L_g denote the Lie derivatives along f and g, respectively. It is assumed that $L_g b(x) \neq 0$ when $b(x) = 0$.

The assumption in the above definition that $L_g b(x) \neq 0$ ensures that the CBF is not subject to a singularity (i.e., $L_g b(x) = 0$) at the boundary of the safe set. As long as this assumption holds, a control u can be determined through (2.13) so as to preserve forward invariance in terms of C.

Recall that the Lie derivative of a function along a vector field captures the change in the value of the function along the vector field (see [26] for detailed descriptions). It is important to note that the expression (2.13) in Definition 2.3 is *linear* in u. This is a crucial point because, as we will see, it leads to computationally efficient solutions for optimization problems based on the CBF method; it is also the reason why we need to assume the dynamics are affine in control, as in (2.11). CBFs as defined in (2.13) are usually called "zeroing CBFs," as the lower bound of the derivative of the function $b(x)$ is limited to 0 when $b(x)$ approaches zero. One can also construct so-called reciprocal CBFs [2], which are basically equivalent to zeroing CBFs. For instance, we may define a reciprocal CBF $B(x) = \frac{1}{b(x)}$, in which case the control should satisfy:

$$L_f B(x) + L_g B(x)u - \frac{\gamma}{B(x)} \leq 0, \tag{2.14}$$

for $\gamma > 0$. If we use $B(x) = \frac{1}{b(x)}$ in the above constraint, we obtain

$$L_f b(x) + L_g b(x)u + \gamma b^3(x) \geq 0, \tag{2.15}$$

which is equivalent to the CBF constraint in (2.13) if we choose the class $\mathcal{K}$ function $\alpha(\cdot)$ as a cubic one. For any other choices of $B(x)$, such as $B(x) = -log(\frac{b(x)}{1+b(x)})$, we can also find a corresponding class $\mathcal{K}$ function that relates a *reciprocal* CBF to a *zeroing* CBF defined in Def. 2.3.

Remark 2.1 Strictly speaking, the CBF in Definition 2.3 should be called a *candidate* CBF because $\alpha(\cdot)$ is not fixed. For simplicity, we omit this qualification throughout the rest of the book based on the premise that we can always find an appropriate CBF for the problems we consider. In fact, in subsequent chapters, we show how to define a class $\mathcal{K}$ function $\alpha(\cdot)$ such that there exists a control $u \in U$ that satisfies (2.13). A CBF is completely defined once $\alpha(\cdot)$ is specified. We may also replace the class $\mathcal{K}$ function in (2.13) by an extended class $\mathcal{K}$ function to ensure robustness to perturbations and compatibility with the Lyapunov-like condition (2.10).

Example 2.2 (Example 2.1 revisited) For the ACC in Example 2.1, consider a different safety constraint expressed as $z \geq 1.8v + l_0$ where the distance between the ego and its preceding vehicle is now limited by a speed-dependent function; the constant $1.8s$ is the typical time-to-collision (or driver reaction time) used in transportation systems (e.g., see [67]). In order to apply (2.13), observe that $L_f b(x) = v_p - v$ and $L_g b(x) = -1.8$, where $b(x) = z - 1.8v - l_0$ is a CBF. Thus, the corresponding CBF constraint in this case is

$$v_p - v + \underbrace{-1.8}_{L_f b(x)} \; \underbrace{u}_{L_g b(x)} + \alpha(z - 1.8v - l_0) \geq 0,$$

where $\alpha(\cdot)$ is a class $\mathcal{K}$ function that is not defined yet (the CBF $b(x)$ in this example is a candidate). The above constraint is a continuous constraint on the ego vehicle state and control. In other words, the state-feedback safe controllers are given as a set:

$$k_{safe}(x) = \{u \in \mathbb{R} : v_p - v - 1.8u + \alpha(z - 1.8v - l_0) \geq 0\}.$$

$\blacksquare$

We conclude this section with an important theorem capturing a crucial property of CBFs known as *forward invariance*:

Theorem 2.1 (Forward invariance of CBFs [2], [21]) *Given a CBF b with the associated set C from (2.12), any Lipschitz continuous controller $u(t)$, $\forall t \geq 0$ that satisfies (2.13) renders the set C forward invariant for (2.11).*

In simple terms, a system with dynamics of the form (2.7) that starts in a safe state x, i.e., $b(x) \geq 0$, is guaranteed to remain safe as long as its control is selected to satisfy (2.13).

2.4   Control Lyapunov Functions

A CBF as defined in Definition 2.3 is a direct extension of the notion of a Control Lyapunov Function (CLF), which is used to enforce stability or state convergence requirements. We also introduce the definition of a CLF here since it is widely used in this book as well.

Definition 2.4 (*Control Lyapunov Function (CLF)* [1]) A continuously differentiable function $V : \mathbb{R}^n \to \mathbb{R}$ is an exponentially stabilizing Control Lyapunov Function (CLF) for system (2.11) if there exist constants $c_1 > 0, c_2 > 0, c_3 > 0$ such that for all $x \in \mathbb{R}^n$, $c_1||x||^2 \leq V(x) \leq c_2||x||^2$,

$$\inf_{u \in U} [L_f V(x) + L_g V(x)u + c_3 V(x)] \leq 0. \tag{2.16}$$

> **Theorem 2.2** ([1]) *Given an exponentially stabilizing CLF V as in Definition 2.4, any Lipschitz continuous controller $u(t)$ that satisfies (2.16) for all $t \geq t_0$ exponentially stabilizes system (2.11) to its origin.*

2.5 Constrained Optimal Control Problem

We now consider a broad class of constrained optimal control problems which, as argued in Chap. 1, are prohibitively difficult to solve in real time. We will see that by defining CBFs corresponding to the constraints, we can replace them by modified ones of the form (2.13), which lead to computationally efficient solutions suitable for real-time applications.

We formally set up a constrained optimal control problem as described next. The problem consists of an objective function, a set of safety requirements expressed as hard constraints on the system state, and a set of constraints on the control.

Objective: Consider the following cost function to be minimized for a system with dynamics (2.11):

$$\min_{u(t)} \int_0^T \mathcal{C}(||u(t)||)dt + p_t||x(T) - K||^2, \tag{2.17}$$

where $|| \cdot ||$ denotes the 2-norm of a vector, $\mathcal{C}(\cdot)$ is a strictly increasing function of its argument, $T > 0$, and $p_t \geq 0$. The second term above is a state convergence requirement such that we wish the terminal state $x(T)$ to reach a desired point $K \in \mathbb{R}^n$. This terminal cost can be easily extended to multiple terminal costs for different state components.

Safety requirements: System (2.11) should always satisfy a safety requirement in the form:

$$b(x(t)) \geq 0, \quad \forall t \in [0, T], \tag{2.18}$$

where $b : \mathbb{R}^n \to \mathbb{R}$ is continuously differentiable.

Control constraints: System (2.11) is subject to control bounds in the form:

$$U := \{u \in \mathbb{R}^q : u_{min} \leq u \leq u_{max}\}. \tag{2.19}$$

where $u_{min}, u_{max} \in \mathbb{R}^q$ and the inequalities are interpreted componentwise.

A control policy for system (2.11) is *feasible* if constraints (2.18) and (2.19) are satisfied at all times. Note that state limitations are particular forms of (2.18) that only depend on the state x. Central to this book is the study of the following problem that we will refer to as Optimal Control Problem (**OCP**):

OCP: Find a feasible control policy (i.e., satisfying (2.18) and (2.19)) for system (2.11) such that the cost (2.17) is minimized. $\square$

Remark 2.2 Before proceeding, we should point out that **OCP** is not as general as the optimal control problem formulated in Chap. 1. First, even though the system dynamics (2.11) are quite general, they are still required to be affine in control. Second, the cost function (2.17) is limited to a dependence on control, and it does not allow state dependence over $[0, T)$, except for a terminal state cost. Moreover, the time horizon T is fixed, excluding, for example, minimum time problems, where minimizing T is part of the objective. Finally, an implicit assumption in (2.18) is that $b(x(t))$ is such that its time derivative explicitly includes the control u; otherwise, $L_g b(x) = 0$ in the second term of (2.13), which prevents us from employing a CBF constraint to specify u. The rest of the book is devoted to addressing all of these limitations and detailing how they can be dealt with.

2.5.1 CBF-Based Approach

In order to make use of CBFs, the first step is to enforce the safety constraint (2.18) by employing a CBF constraint through (2.13). As already explained, the latter is a sufficient condition for the former to be satisfied, so it generally involves some conservativeness. On the other hand, it provides an inequality that is linear in the control u, hence offering a simple way for specifying the control that ensures satisfaction of (2.18). Moreover, it possesses the forward invariance property of Theorem 2.1. The second step is to employ a CLF, as defined in Definition 2.4, to implement the desired state convergence expressed as a terminal state cost in (2.17). Using this CBF/CLF approach, which for simplicity we refer to as the *CBF-based approach*, the original problem **OCP** becomes **OCP-CBF**:

$$\min_{u(t), \delta_r(t)} \int_0^T [\mathcal{C}(\|u(t)\|) + p_t \delta_r^2(t)] dt$$

subject to

$$L_f b(x) + L_g b(x)u + \alpha(b(x)) \geq 0,$$
$$L_f V(x) + L_g V(x)u + \epsilon V(x) \leq \delta_r,$$
$$u_{min} \leq u \leq u_{max},$$

$$(2.20)$$

where $V(x) = (x - K)^T P(x - K)$, P is positive definite, and δ_r is a relaxation variable that allows us to treat the terminal state constraint in (2.17) as a "soft" CLF-based constraint on the control by seeking values of $\delta_r(t)$ to prevent it from conflicting with the CBF constraint. Naturally, Problem **OCP-CBF** (2.20) is not equivalent to **OCP** since the constraints above are only sufficient conditions for the constraints in (2.18). However, as described next, solving **OCP-CBF** (2.20) is a much simpler computational task and guarantees the satisfaction of all safety and control constraints, assuming that feasible solutions indeed exist. Therefore, the CBF-based formulation offers a tradeoff between computational complexity and some potential conservativeness in the solutions it provides.

It is of course possible to derive explicit analytical solutions to optimization problems of the form **OCP-CBF** (2.20), especially when certain assumptions apply (e.g., see [4]). However, this is not possible in general, especially when the number of safety constraints is large. Alternatively, by treating the state x as an implicit decision variable, we can formulate a receding horizon optimization problem. This, however, most likely results in CBF constraints that are nonlinear in x and u (especially for nonlinear dynamics and constraints), which makes the overall optimization problem hard to solve.

This motivates seeking a general-purpose procedure to ensure that **OCP-CBF** (2.20) can always be efficiently solved, at least numerically, with any desired degree of accuracy. To do so, following [2], we can discretize the time interval uniformly, and fix the state x of the optimization problem within each time interval of length Δt to the value at the beginning of the interval. Note that the CBF and the CLF constraints are then linear in control. This allows us to formulate a sequence of convex optimization problems (in particular, quadratic in the case of **OCP-CBF** (2.20) if C is a quadratic function), which can be efficiently solved.

We conclude this section by describing in detail the procedure for obtaining a complete solution for **OCP-CBF** (2.20) through time discretization. Specifically, we partition the time interval $[0, T]$ into a set of equal time intervals $\{[0, \Delta t), [\Delta t, 2\Delta t), \ldots\}$, with $\Delta t > 0$. In each interval $[\omega\Delta t, (\omega + 1)\Delta t)$, $\omega = 0, 1, 2, \ldots$, we keep the state constant at its value at the beginning of the interval and also assume that the control remains constant. We then reformulate **OCP-CBF** (2.20) as a sequence of point-wise optimization problems. Specifically, at $t = \omega\Delta t$, $\omega = 0, 1, 2, \ldots$, we solve

$$
\min_{u(t), \delta_r(t)} C(\|u(t)\|) + p_t \delta_r^2(t)
$$
$$
\text{s.t. } L_f b(x) + L_g b(x)u + \alpha(b(x)) \geq 0,
$$
$$
L_f V(x) + L_g V(x)u + \epsilon V(x) \leq \delta_r, \tag{2.21}
$$
$$
u_{min} \leq u \leq u_{max},
$$

where $u(t)$, $\delta(t)$ with t fixed at $t = \omega\Delta t$ are the decision variables of a point-wise Quadratic Program (QP) if $C(\|u(t)\|)$ is quadratic in u. By solving (2.21), we obtain the optimal values $u^*(t)$, $\delta^*(t)$. We then integrate (2.11) with control $u^*(t)$ kept constant during $[\omega\Delta t, (\omega + 1)\Delta t)$. We emphasize again that this represents an approximate solution of the original optimal control problem (2.17), since the optimization process above is performed point-wise and the CBF constraints are only sufficient conditions for the original safety constraints in (2.17). The last two chapters of this book include challenging constrained optimization problems for autonomous systems that arise in transportation networks and robotics, where the CBF-based method is shown to yield real-time solutions that guarantee safety and can be shown to be near-optimal.

Remark 2.3 (*Comparison between the CBF-based approach and Model Predictive Control (MPC)*) In the MPC approach [51], the optimization is defined over a receding horizon.

In contrast, the CBF method is myopic in the sense that the optimization is carried out over a single step at a time. It follows that the MPC optimization process is more likely to yield feasible solutions. However, it is significantly more demanding computationally as the optimization is performed over state variables, in addition to control variables, and the system dynamics are used as constraints. The optimization problem in the MPC approach is also, in general, a nonlinear program.

2.5.2   Feasibility in the CBF-Based Approach

The CBF-based optimization problem **OCP-CBF** (2.20) based on the original **OCP** may not have a solution. This would definitely be the case if the original problem itself is not feasible. Here we focus on the interesting cases in which **OCP** is feasible and **OCP-CBF** (2.20) is not. This can occur since we replaced the original safety constraints (2.18) with CBF constraints. Feasibility depends, for instance, on the possible conflict between the CBF constraints and the control bounds included in the formulation of **OCP-CBF** (2.20). This, in turn, depends on the choice of class $\mathcal{K}$ functions that we select in (2.13).

The discretization QP-based solution approach presented in (2.21), can cause even more feasibility problems. In particular, a critical issue in this approach is the proper choice of Δt, which is related to the *inter-sampling effect*: we need to make sure that the CBF constraint is satisfied within *every* time interval $[\omega \Delta t, (\omega + 1)\Delta t), \omega = 0, 1, 2, \ldots$, i.e., at every (continuous) time instant in every interval. Otherwise, the forward invariance property in Theorem 2.1 may be violated. It is, however, possible that while a QP has a feasible solution at some time step, the next step's QP fails to have a feasible solution unless this can be explicitly ensured. There are several ways to address the inter-sampling effect, including proper selection of the Lipschitz constant, as in [100]. This also opens up the question of replacing the time-driven sequence of QPs in (2.21) by an *event-driven* mechanism, as, for example, done in [63] using a procedure inspired from event-triggered control for Lyapunov functions [58].

2.5.3   Time-Driven and Event-Driven CBFs

Most existing work to date based on the QP approach (2.21) uses a uniform time discretization, i.e., time-driven CBFs. This is, however, not necessary if one can define proper events such that the next QP is solved only when the conditions defining any such event are satisfied. Therefore, a key challenge is to adapt this process to an event-driven mechanism so as to determine the next time when a QP needs to be solved in order to guarantee safety. In Chap. 7, we will present a robust event-driven CBF approach that also works for systems with unknown dynamics. This will eliminate the need of selecting a sampling interval length

Δt and instead define appropriate *events* whose occurrence determines the next time that an instance of the convex optimization problem needs to be solved.

The next few chapters systematically address the feasibility problem in optimization with CBFs. We will revisit the limitations presented earlier in Remark 2.2 along with the question of determining an appropriate class $\mathcal{K}$ function α in Definition 2.3 that ensures the existence of a CBF and, ultimately, the existence of solutions to the optimization problem **OCP-CBF** (2.20). We begin in Chap. 3 with the issue that arises when the control does not explicitly show up in the corresponding first derivative of a CBF $b(x)$ associated with a safety constraint $b(x) \geq 0$, This motivates extending CBFs to *High-Order* CBFs (HOCBFs).

High Order Control Barrier Functions

3

3.1   High Order Barrier Functions

In Definition 2.3, it is easy to see that a controller ensuring the safety condition can only be found if $L_g b(x) \neq 0$. This assumption, which limits the application of the approach from the previous chapter, is illustrated in the example below.

Example 3.1 (Example 2.1 revisited) For the ACC system, recall that we require the distance $z(t)$ between the ego vehicle and its immediately preceding vehicle to be greater than a constant $\delta > 0$ for all the times, i.e.,

$$z(t) \geq \delta, \forall \ t \geq t_0. \tag{3.1}$$

Let $x(t) := (v(t), z(t))$ and $b(x(t)) = z(t) - \delta$. With $\alpha(\cdot)$ from Definition 2.3 chosen as the identity function, according to (2.13), in order to ensure safety, we need to have:

$$\underbrace{v_p - v(t)}_{L_f b(x(t))} + \underbrace{0}_{L_g b(x(t))} \times u(t) + \underbrace{z(t) - \delta}_{b(x(t))} \geq 0. \tag{3.2}$$

Note that $L_g b(x(t)) = 0$ in (3.2), i.e., $u(t)$ does not show up. Therefore, this condition provides no information on a proper control to be applied and we cannot use this CBF to formulate an optimization problem as described at the end of Chap. 2. ∎

    Motivated by this simple example, we begin by defining the concept of *relative degree* for a given function.

© The Author(s), under exclusive license to Springer Nature Switzerland AG 2023

W. Xiao et al., *Safe Autonomy with Control Barrier Functions*,

Synthesis Lectures on Computer Science,

https://doi.org/10.1007/978-3-031-27576-0_3

Definition 3.1 *(Relative degree* [26]*)* The relative degree of a (sufficiently) differentiable function $b : \mathbb{R}^n \to \mathbb{R}$ with respect to system (2.11) is the number of times we need to differentiate it along the dynamics of (2.11) until any component of the control $\boldsymbol{u}$ explicitly shows.

Since the function b is used to define a constraint $b(\boldsymbol{x}) \geq 0$, we will also refer to the relative degree of b as the relative degree of the constraint.

For the sake of generality, we consider a time-varying function to define an invariant set for system (2.4). The proposed time-varying High Order Barrier Functions apply to time-varying systems as well. For a mth order differentiable function $b : \mathbb{R}^n \times [t_0, \infty) \to \mathbb{R}$, we recursively define a sequence of functions $\psi_i : \mathbb{R}^n \times [t_0, \infty) \to \mathbb{R}, i \in \{1, \ldots, m\}$ in the form:

$$\psi_i(\boldsymbol{x}, t) = \dot{\psi}_{i-1}(\boldsymbol{x}, t) + \alpha_i(\psi_{i-1}(\boldsymbol{x}, t)), \quad i \in \{1, \ldots, m\}, \tag{3.3}$$

where $\alpha_i(\cdot), i \in \{1, \ldots, m\}$ denote class $\mathcal{K}$ functions of their argument and $\psi_0(\boldsymbol{x}, t) = b(\boldsymbol{x}, t)$.

We further define a sequence of sets $C_i(t), i \in \{1, \ldots, m\}$ associated with (3.3) in the form:

$$C_i(t) = \{\boldsymbol{x} \in \mathbb{R}^n : \psi_{i-1}(\boldsymbol{x}, t) \geq 0\}, \quad i \in \{1, \ldots, m\}. \tag{3.4}$$

Definition 3.2 *(High Order Barrier Function (HOBF))* Let $C_i(t), i \in \{1, \ldots, m\}$ be defined by (3.4) and $\psi_i(\boldsymbol{x}, t), i \in \{1, \ldots, m\}$ be defined by (3.3). A function $b : \mathbb{R}^n \times [t_0, \infty) \to \mathbb{R}$ is a High Order Barrier Function (HOBF) for (2.4) if it is m^{th} order differentiable and there exist differentiable class $\mathcal{K}$ functions $\alpha_i, i \in \{1, \ldots m\}$ s.t.

$$\psi_m(\boldsymbol{x}, t) \geq 0 \tag{3.5}$$

for all $(\boldsymbol{x}, t) \in C_1(t) \cap, \ldots, \cap C_m(t) \times [t_0, \infty)$.

Theorem 3.1 ([71]) *The set $C_1(t) \cap, \ldots, \cap C_m(t)$ is forward invariant for system (2.4) if $b(\boldsymbol{x}, t)$ is a HOBF.*

Remark 3.1 The sets $C_i(t), i \in \{1, \ldots, m\}$ should have a non-empty intersection at t_0 in order to satisfy the forward invariance condition starting from t_0 in Theorem 3.1. If $b(\boldsymbol{x}(t_0), t_0) \geq 0$, we can always choose proper class $\mathcal{K}$ functions $\alpha_i(\cdot), i \in \{1, \ldots, m\}$ to make $\psi_i(\boldsymbol{x}(t_0), t_0) \geq 0, \forall i \in \{1, \ldots, m - 1\}$. There are some extreme cases, however, when this is not possible. For example, if $\psi_0(x(t_0), t_0) = 0$ and $\dot{\psi}_0(x(t_0), t_0) < 0$, then $\psi_1(x(t_0), t_0)$ is always negative no matter how we choose $\alpha_1(\cdot)$. Similarly, if $\psi_0(x(t_0), t_0) = 0$, $\dot{\psi}_0(x(t_0), t_0) = 0$ and $\dot{\psi}_1(x(t_0), t_0) < 0$, $\psi_2(x(t_0), t_0)$ is also always negative, etc. To deal with such extreme cases (as with the case when $b(x(t_0), t_0) < 0$), we would need a feasibility enforcement method, which depends on the particular problem of interest. We shall henceforth not consider extreme cases of this type.

3.2   High Order Control Barrier Functions (HOCBFs)

We are now ready to provide an extension of the CBF from Definition 2.3.

Definition 3.3 *(High Order Control Barrier Function (HOCBF))* Let $C_i(t), i \in \{1, \ldots, m\}$ be defined by (3.4) and $\psi_i(x, t), i \in \{1, \ldots, m\}$ be defined by (3.3). A function $b : \mathbb{R}^n \times [t_0, \infty) \to \mathbb{R}$ is a High Order Control Barrier Function (HOCBF) of relative degree m for system (2.11) if there exist differentiable class $\mathcal{K}$ functions $\alpha_i, i \in \{1, \ldots, m\}$ s. t.

$$\sup_{u \in U}[\mathcal{L}_f^m b(x, t) + \mathcal{L}_g \mathcal{L}_f^{m-1} b(x, t)u + \frac{\partial^m b(x, t)}{\partial t^m}$$

$$+ O(b(x, t)) + \alpha_m(\psi_{m-1}(x, t))] \geq 0, \tag{3.6}$$

for all $(x, t) \in C_1(t) \cap, \ldots, \cap C_m(t) \times [t_0, \infty)$. $\mathcal{L}_f, \mathcal{L}_g$ denote the partial Lie derivatives w.r.t. x along f and g, respectively, and $O(\cdot)$ is given by

$$O(b(x, t)) = \sum_{i=1}^{m-1} \mathcal{L}_f^i(\alpha_{m-i} \circ \psi_{m-i-1})(x, t) + \frac{\partial^i(\alpha_{m-i} \circ \psi_{m-i-1})(x, t)}{\partial t^i}$$

Moreover, it is assumed that $\mathcal{L}_g \mathcal{L}_f^{m-1} b(x, t) \neq 0$ when $b(x, t) = 0$.

Similar to Definition 2.3, a HOCBF is defined when $\alpha_i(\cdot), i \in \{1, \ldots, m\}$ are specified. Subsequent chapters are dedicated to finding such functions. Given a HOCBF $b(x, t)$, we define the set of controls that satisfy:

$$K_{hocbf}(x, t) = \{u \in U : \mathcal{L}_f^m b(x, t) + \mathcal{L}_g \mathcal{L}_f^{m-1} b(x, t)u$$

$$+ \frac{\partial^m b(x, t)}{\partial t^m} + O(b(x, t)) + \alpha_m(\psi_{m-1}(x, t)) \geq 0\}. \tag{3.7}$$

> **Theorem 3.2** ([71]) *Given a HOCBF $b(\boldsymbol{x}, t)$ from Definition 3.3 with the associated sets $C_i(t), i \in \{1, \ldots, m\}$ defined by (3.4), if $\boldsymbol{x}(t_0) \in C_1(t_0) \cap, \ldots, \cap C_m(t_0)$, then any Lipschitz continuous controller $\boldsymbol{u}(t) \in K_{hocbf}(\boldsymbol{x}(t), t), \forall t \geq t_0$ renders the set $C_1(t) \cap, \ldots, \cap C_m(t)$ forward invariant for system (2.11).*

Remark 3.2 The general, time-varying HOCBF from Definition 3.3 can be used for general, time-varying constraints (e.g., signal temporal logic specifications [31]) and systems. However, many problems of interest have time-invariant system dynamics and constraints. Therefore, in the rest of this book, we focus on such time-invariant versions for simplicity. In other words, the time-invariant HOCBF constraint corresponding to (3.6) is in the form:

$$L_f^m b(\boldsymbol{x}) + L_g L_f^{m-1} b(\boldsymbol{x})\boldsymbol{u} + O(b(\boldsymbol{x})) + \alpha_m(\psi_{m-1}(\boldsymbol{x})) \geq 0, \tag{3.8}$$

where $O(b(\boldsymbol{x})) = \sum_{i=1}^{m-1} L_f^i(\alpha_{m-i} \circ \psi_{m-i-1})(\boldsymbol{x})$

Remark 3.3 *(Relationship between time-invariant HOCBF and exponential CBF in [43])* In Definition 3.2, if we set class $\mathcal{K}$ functions $\alpha_1, \alpha_2 \ldots \alpha_m$ to be linear functions with positive coefficients, then we can get exactly the same formulation as in [43] that is obtained through input-output linearization. i.e.,

$$\psi_i(\boldsymbol{x}) = \dot{\psi}_{i-1}(\boldsymbol{x}) + k_i \psi_{i-1}(\boldsymbol{x}), i \in \{1, \ldots, m\}, \tag{3.9}$$

where $k_i > 0, i \in \{1, \ldots, m\}$. Therefore, the time-invariant version HOCBF we have defined is a generalization of the exponential CBF introduced in [43].

Example 3.2 (Example 2.1 revisited) For the ACC example, the relative degree of the constraint from (3.1) is 2. Therefore, we need a HOCBF with $m = 2$. We choose quadratic class $\mathcal{K}$ functions for both $\alpha_1(\cdot)$ and $\alpha_2(\cdot)$, i.e., $\alpha_1(b(\boldsymbol{x}(t))) = b^2(\boldsymbol{x}(t))$ and $\alpha_2(\psi_1(\boldsymbol{x}(t))) = \psi_1^2(\boldsymbol{x}(t))$. In order for $b(\boldsymbol{x}(t)) := z(t) - \delta$ to be a HOCBF for (2.11), a control input $u(t)$ should satisfy

$$\begin{aligned}
L_f^2 b(\boldsymbol{x}(t)) + L_g L_f b(\boldsymbol{x}(t))u(t) + 2b(\boldsymbol{x}(t))L_f b(\boldsymbol{x}(t)) \\
+ (L_f b(\boldsymbol{x}(t)))^2 + 2b^2(\boldsymbol{x}(t))L_f b(\boldsymbol{x}(t)) + b^4(\boldsymbol{x}(t)) \geq 0.
\end{aligned} \tag{3.10}$$

Note that $L_g L_f b(\boldsymbol{x}(t)) \neq 0$ in (3.10) and the initial conditions are $b(\boldsymbol{x}(t_0)) \geq 0$ and $\dot{b}(\boldsymbol{x}(t_0)) + b^2(\boldsymbol{x}(t_0)) \geq 0$. ∎

In conclusion, to formulate a HOCBF-based optimization problem that guarantees safety for systems with arbitrary relative degree, we simply replace the CBF constraint in **OCP-CBF** (2.20) by the HOCBF constraint (3.8).

3.3   ACC Using HOCBFs

We are now ready to complete the ACC case study introduced in Example 3.1. For the dynamics given by (2.8), we consider a cost $J(u) = \int_{t_0}^{t_f} u^2(t)dt$ and we require the vehicle to achieve a desired speed $v_d = 24m/s$. For this purpose, we define a CLF $V(x) = (v - v_d)^2$ (see Definition 2.4).

We consider a control constraint $-0.4g \leq u(t) \leq 0.4g$, $g = 9.81m/s^2$. Note that the relative degree of (2.5) is $m = 2$, and we define two different HOCBFs as in Definition 3.3 for $b(x) = z - \delta$ by choosing linear and quadratic class $\mathcal{K}$ functions for both $\alpha_1(\cdot), \alpha_2(\cdot)$ in (4.1) with constants $p_1 = p_2 = p > 0$. In particular, we set $p = 1, 0.02$ for linear and quadratic class $\mathcal{K}$ functions, respectively. We present the forward invariance of the set $C_1 \cap C_2$, where $C_1 := \{x(t) : b(x(t)) \geq 0\}$ and $C_2 := \{x(t) : \psi_1(x(t)) \geq 0\}$ in Fig. 3.1. All simulations were conducted in MATLAB. We used quadprog to solve the QPs and ode45 to integrate the dynamics. The remaining parameters in this problem are $v(t_0) = 20m/s$, $z(t_0) = 100m$, $v_0(t) = 13.89m/s$, $\delta = 10m$, $\Delta t = 0.1s$. Additional details can be found in [71].

3.4   HOCBFs for Systems with Multiple Inputs

In this section, we consider systems with multiple control inputs where all or a desired subset of the control components needs to be used. In this case, we need to guarantee (safety) constraint satisfaction using HOCBFs such that this desired subset shows up in the corresponding HOCBF constraint (3.8).

Suppose there is a safety requirement $b(x) \geq 0$ for system (2.11). If we enforce this constraint using a HOCBF, we first need to determine the relative degree of $b(x)$. As sys-

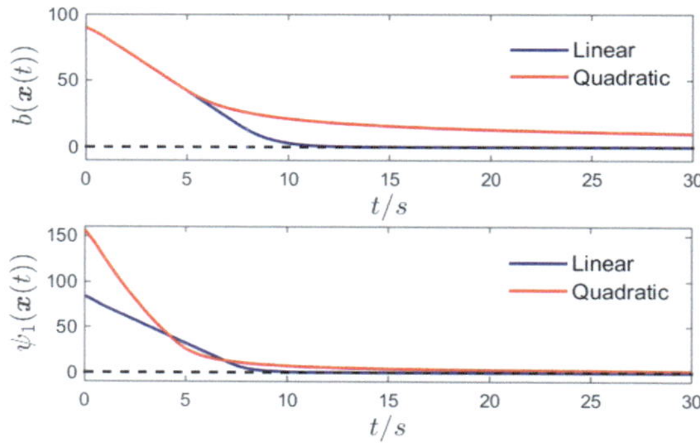

Fig. 3.1 The functions $b(x(t))$ and $\psi_1(x(t))$ from Example 3.1 for linear ($p = 1$) and quadratic ($p = 0.02$) class $\mathcal{K}$ functions, respectively. $b(x(t)) \geq 0$ and $\psi_1(x(t)) \geq 0$ imply the forward invariance of $C_1 \cap C_2$

tem (2.11) has multiple control inputs, one option is to consider the relative degree as the minimum number of times that we differentiate $b(x)$ along the dynamics (2.11) until any component of the control vector shows up in the corresponding derivative. However, this may reduce the system performance, since it implies that only a limited set of control components can be used to guarantee constraint satisfaction. As an example, to make an autonomous vehicle satisfy a safety constraint with respect to a preceding vehicle using HOCBFs, we may require the ego vehicle to follow the preceding vehicle or to overtake it. In the former case, steering wheel control is not necessary to show up in the HOCBF constraint (3.8). However, in the latter case, we wish that both the acceleration control and steering wheel control show up in the HOCBF constraint (3.8). This can improve the mobility of an autonomous vehicle compared to the former case, as well as significantly improve the feasibility of the resulting HOCBF-based QPs in (2.21) when control bounds are involved.

In this section, we focus on the case where we wish *all* the components of the control vector to show up in the HOCBF constraint (3.8). We can always fix the control (e.g., setting it to 0) of undesired control components in the HOCBF constraint (3.8) if we wish to guarantee safety using *some* desired control components. In order to achieve this, we define the relative degree set $S \subset \mathbb{N}$ of a function $b : \mathbb{R}^n \to \mathbb{R}$ as follows:

> **Definition 3.4** *(Relative degree set)* The relative degree set $S \subset \mathbb{N}$ of a function $b : \mathbb{R}^n \to \mathbb{R}$ with respect to system (2.11) is defined by the ordered set of numbers of times we need to differentiate b along system (2.11) until each component of the control vector u first shows in the corresponding derivative.

This definition is illustrated in the following example along with some issues which we will resolve through the use of integral HOCBFs in the next section.

Example 3.3 Consider a simplified unicycle model of a robot that has the form:

$$\dot{x} = v \cos \theta, \quad \dot{y} = v \sin \theta, \quad \dot{v} = \frac{u_2}{M}, \quad \dot{\theta} = \phi, \quad \dot{\phi} = u_1, \tag{3.11}$$

where $(x, y) \in \mathbb{R}^2$ denotes the 2-D location of the system, $v \in \mathbb{R}$ denotes the linear speed, $\theta \in \mathbb{R}$ is the heading angle, $\phi \in \mathbb{R}$ denotes the rotation speed, $M > 0$ is the mass of the system, and $u_1 \in \mathbb{R}, u_2 \in \mathbb{R}$ stand for the angular acceleration and driven force (control inputs), respectively.

Suppose we have a state constraint:

$$\sqrt{(x - x_0)^2 + (y - y_0)^2} \geq r, \tag{3.12}$$

where $(x_0, y_0) \in \mathbb{R}^2$, and $r > 0$. With $b(x) := \sqrt{(x - x_0)^2 + (y - y_0)^2} - r$, we can see that the relative degree of b with respect to u_1 is 3, and the relative degree of b with respect to u_2 is 2. Therefore, the relative degree set of $b(x)$ is $S = \{3, 2\}$.

When defining a HOCBF (taking class $\mathcal{K}$ functions α_1, α_2 to be linear) for the constraint (2.5) with respect to u_2, i.e., the relative degree $m = 2$ in Definition 3.3, we have

$$L_f^2 b(x) + L_{g_2} L_f b(x) u_2 + 2L_f b(x) + b(x) \geq 0, \tag{3.13}$$

where $g_2 = [0, 0, \frac{1}{M}, 0, 0]^T$ is the second column of $g(x)$ in (3.11) when it is written in vector form.

We can also define a HOCBF (taking class $\mathcal{K}$ functions $\alpha_1, \alpha_2, \alpha_3$ to be linear) for the constraint (3.12) with respect to u_1, i.e., the relative degree $m = 3$ in Definition 3.3, and we have

$$\begin{aligned}
L_f^3 b(x) + L_{g_1} L_f^2 b(x) u_1 + L_f [L_{g_2} L_f b(x)] u_2 + L_{g_2} L_f b(x) \dot{u}_2 \\
+ 3L_f^2 b(x) + 3L_{g_2} L_f b(x) u_2 + 3L_f b(x) + b(x) \geq 0,
\end{aligned} \tag{3.14}$$

where $g_1 = [0, 0, 0, 0, 1]^T$ is the first column of $g(x)$ in (3.11) when it is written in vector form. Note that the derivative of u_2 is included in the above.

There is only one control input in the HOCBF constraint (3.13). The safety constraint (3.12) can only be guaranteed using u_2 (not u_1) and the feasibility of the QPs (2.21) is also impaired by control bounds as in (2.19). In other words, a robot can only use linear deceleration to avoid the obstacle specified by the constraint (3.12). On the other hand, (3.14) includes the derivative of u_2, and this introduces an additional problem (i.e., how to choose $\dot{u}_2$) in the HOCBF-based QP. ∎

Before defining integral HOCBFs, we introduce some additional notation that will facilitate the analysis that follows. There may be some control components of u that are differentiated at least once in the HOCBF constraint, and we define the ordered index set of those differentiated control components as S_d. The cardinality of S_d is denoted by $N_d \in \mathbb{N}$, $N_d < q$ where q is the dimension of u. Let u_n denote the control vector that includes only those control components whose derivatives are never present in the corresponding HOCBF constraint (e.g., $u_n = u_1$ in (3.14)), and let U_n denote the control constraint set, defined in (2.19), corresponding to u_n. Let $g_n : \mathbb{R}^n \to \mathbb{R}^{n \times (q - N_d)}$ denote a matrix that is composed of the columns of the matrix $g(x)$ in (2.11) corresponding to each control component in u_n.

Let $\overline{m}$ denote the maximum relative degree of $b(x) \geq 0$ with respect to (2.11), i.e., $\overline{m} = \max_{k \in S} k$. Then, we define $b(x)$ to be a HOCBF with relative degree $\overline{m}$. In order to deal with the derivatives of the control components, we define auxiliary dynamics for each $u_j, j \in S_d$ that is differentiated m_j times, $1 \leq m_j < \overline{m}$:

$$\dot{u}_j = f_j(u_j) + g_j(u_j) v_j, \tag{3.15}$$

where $u_j = (u_{j,1}, u_{j,2}, \ldots, u_{j,m_j}) \in \mathbb{R}^{m_j}$ is the corresponding auxiliary state, and $u_{j,1} = u_j$, $f_j : \mathbb{R}^{m_j} \to \mathbb{R}^{m_j}$, $g_j : \mathbb{R}^{m_j} \to \mathbb{R}^{m_j}$, and $v_j \in \mathbb{R}$ is a new control for the auxiliary dynamics (3.15) corresponding to u_j. The relative degree of u_j (now a state variable) with respect to (3.15) is m_j, and $\boldsymbol{u}_j$ can be initialized to any vector as long as u_j strictly satisfies its control bound in (2.19). Although f_j, g_j may be arbitrarily selected, for simplicity, we may define (3.15) in linear form, and initialize $u_{j,k}$, $k \in \{2, \ldots, m_j\}$ to 0. Further, let $\boldsymbol{v} \in \mathbb{R}^{N_d}$ be the concatenation of v_j, $\forall j \in S_d$, and let $\boldsymbol{u}_a$ be the concatenation of $\boldsymbol{u}_j$, $\forall j \in S_d$. Thus, u_j, $j \in S_d$, unlike $\boldsymbol{u}_n$, contains all control components whose derivatives appear at least once in the HOCBF constraint.

Combining the auxiliary dynamics (3.15), we get the HOCBF constraint (relative degree $\overline{m}$) enforcing $b(\boldsymbol{x}) \geq 0$:

$$L_f^{\overline{m}} b(\boldsymbol{x}) + [L_{g_n} L_f^{\overline{m}-1} b(\boldsymbol{x})] \boldsymbol{u}_n + O(b(\boldsymbol{x}), \boldsymbol{u}_a, \boldsymbol{v})$$
$$+ \alpha_{\overline{m}}(\psi_{\overline{m}-1}(\boldsymbol{x}, \boldsymbol{u}_a)) \geq 0, \tag{3.16}$$

where $O(\cdot)$ is defined as in (3.8), but also includes the derivatives of u_j, $\forall j \in S_d$, i.e., $u_j^{(1)}, \ldots, u_j^{(m_j)}$, and $u_j^{(m_j)}$ denotes the m_j^{th} derivative of u_j.

In order to apply the CBF-based QP approach in (2.21) so as to guarantee (safety) constraint satisfaction, we can take $\boldsymbol{u}_n$, v_j, $\forall j \in S_d$ instead of $\boldsymbol{u}$ as the decision variables in each QP. After solving the QP, we obtain the optimal $\boldsymbol{u}_n$, v_j, $\forall j \in S_d$ for each time interval, and the controls u_j, $j \in S_d$ are obtained by solving (3.15). Since this is done by integration, we call this *integral control*. We refer to the resulting HOCBF in (3.16) as an integral HOCBF (iHOCBF), and note that it is a class of integral control barrier functions [3].

As in (2.19), we have control bounds for each u_j, $j \in S_d$:

$$u_{j,min} \leq u_j \leq u_{j,max}, \tag{3.17}$$

where $u_{j,min} \in \mathbb{R}$, $u_{j,max} \in \mathbb{R}$ denote the minimum and maximum control bounds, respectively. In order to guarantee the above control bound (3.17) under the auxiliary dynamics (3.15), we define two HOCBFs for each u_j, $j \in S_d$ to map the bound from u_j to v_j. Letting $b_{j,min}(\boldsymbol{u}_j) = u_j - u_{j,min}$ and $b_{j,max}(\boldsymbol{u}_j) = u_{j,max} - u_j$, we can then define the set of the auxiliary control $\boldsymbol{v}$ that enforces (3.17):

$$U_a(\boldsymbol{u}_a) = \{ \boldsymbol{v} \in \mathbb{R}^{N_d} : L_{f_j}^{m_j} b_{j,min}(\boldsymbol{u}_j) + O(b_{j,min}(\boldsymbol{u}_j))$$
$$+ [L_{g_j} L_{f_j}^{m_j-1} b_{j,min}(\boldsymbol{u}_j)] v_j + \alpha_{m_j}(\psi_{m_j-1}(\boldsymbol{u}_j)) \geq 0,$$
$$L_{f_j}^{m_j} b_{j,max}(\boldsymbol{u}_j) + [L_{g_j} L_{f_j}^{m_j-1} b_{j,max}(\boldsymbol{u}_j)] v_j$$
$$+ O(b_{j,max}(\boldsymbol{u}_j)) + \alpha_{m_j}(\psi_{m_j-1}(\boldsymbol{u}_j)) \geq 0, \forall j \in S_d \}. \tag{3.18}$$

Through the equations above, the states of the auxiliary dynamics (3.15) are strictly bounded for each v_j, $j \in S_d$. This property is captured by the invariant sets defined as in (3.4):

Definition 3.5 *(Integral HOCBF)* Let $C_1, C_2, \ldots, C_{\overline{m}}$ be defined as in (3.4) and $\psi_0(\boldsymbol{x}, \boldsymbol{u}_a), \psi_1(\boldsymbol{x}, \boldsymbol{u}_a), \ldots, \psi_{\overline{m}}(\boldsymbol{x}, \boldsymbol{u}_a)$ be defined as in (3.3). A function $b : \mathbb{R}^n \to \mathbb{R}$ is an integral HOCBF (iHOCBF) of relative degree $\overline{m}$ for system (2.11) if there exist differentiable class $\mathcal{K}$ functions $\alpha_1, \alpha_2, \ldots, \alpha_{\overline{m}}$ and auxiliary dynamics (3.15) such that

$$
\sup_{\boldsymbol{u}_n \in U_n, \boldsymbol{v} \in U_a(\boldsymbol{u}_c)} \; [L_f^{\overline{m}} b(\boldsymbol{x}) + [L_{g_n} L_f^{\overline{m}-1} b(\boldsymbol{x})] \boldsymbol{u}_n
$$
$$
+ O(b(\boldsymbol{x}), \boldsymbol{u}_a, \boldsymbol{v}) + \alpha_{\overline{m}}(\psi_{\overline{m}-1}(\boldsymbol{x}, \boldsymbol{u}_a))] \geq 0,
\tag{3.19}
$$

for all $(\boldsymbol{x}, \boldsymbol{u}_a) \in C_1 \cap C_2 \cap, \ldots, \cap C_{\overline{m}}$.

Note that $u_j, \forall j \in S_d$ and their derivatives become state variables with the auxiliary dynamics (3.15). All control inputs $\boldsymbol{u}_n, \boldsymbol{v}$ are in linear forms if we fix $\boldsymbol{x}, \boldsymbol{u}_j, \forall j \in S_d$. Similar to Theorem 3.2, we also have the following:

Theorem 3.3 ([80]) *Given an iHOCBF $b(\boldsymbol{x})$ from Definition 3.5 with the associated sets $C_1, C_2, \ldots, C_{\overline{m}}$ defined as in (3.4), if $(\boldsymbol{x}(t_0), \boldsymbol{u}_a(t_0)) \in C_1 \cap C_2 \cap, \ldots, \cap C_{\overline{m}}$, then any continuously differentiable controller $(\boldsymbol{u}_n(t), \boldsymbol{v}(t)) \in U_n \times U_a(\boldsymbol{u}_a), \forall t \geq t_0$ that satisfies the constraint in (3.19) renders the set $C_1 \cap C_2 \cap, \ldots, \cap C_{\overline{m}}$ forward invariant for systems (2.11), (3.15). Moreover, the control $\boldsymbol{u}_n$ and integral controls $u_j, \forall j \in S_d$ render C_1 forward invariant for system (2.11).*

Example 3.4 (Example 3.3 revisited.) For the unicycle introduced above, the minimum and maximum relative degrees of (3.12) are 2 and 3, respectively. If we define a HOCBF with relative degree 3 to enforce (3.12), we will have the derivative of u_2 in the corresponding HOCBF constraint. Following the above process, we may define the following simple auxiliary dynamics for u_2: $\dot{u}_2 = v$. Then, the HOCBF (taking class $\mathcal{K}$ functions $\alpha_1, \alpha_2, \alpha_3$ as linear functions) constraint (3.19) for (3.12) is

$$
L_f^3 b(\boldsymbol{x}(t)) + L_{g_1} L_f^2 b(\boldsymbol{x}(t)) u_1(t) + L_f [L_{g_2} L_f b(\boldsymbol{x}(t))] u_2(t)
$$
$$
+ L_{g_2} L_f b(\boldsymbol{x}(t)) v + 3 L_f^2 b(\boldsymbol{x}(t)) + 3 L_{g_2} L_f b(\boldsymbol{x}(t)) u_2(t)
\tag{3.20}
$$
$$
+ 3 L_f b(\boldsymbol{x}(t)) + b(\boldsymbol{x}(t)) \geq 0,
$$

where u_2 in the above is a state variable instead of a decision variable (control) in the QP. The resulting u_2 applied to system (3.11) is obtained through the integration of the auxiliary dynamics $\dot{u}_2 = v$. Meanwhile, the control bound for v is obtained through the two CBFs in (3.18). ∎

Remark 3.4 (HOCBFs for non-affine control systems) Suppose we have general nonlinear systems (2.7) instead of affine control systems (2.11). Then, in order to apply the CBF-based optimization approach introduced at the end of Chap. 2, we can simply augment (2.7) to the following form:

$$\dot{x} = f(x, u),$$
$$\dot{u} = f_a(u) + g_a(u)v, \tag{3.21}$$

where $f_a : \mathbb{R}^q \to \mathbb{R}^q$, $g_a : \mathbb{R}^q \to \mathbb{R}^{q \times q}$, and $v \in \mathbb{R}^q$ is the auxiliary control as in (3.15). Then, we can define iHOCBFs as shown above to guarantee safety for a general nonlinear system (2.7).

3.5   Robot Control Using iHOCBFs

We now complete Example 3.3. For the robot with the unicycle model as in (3.11), we consider the problem of determining a control law that minimizes

$$J(u(t)) = \int_{t_0}^{t_f} \left[u_1^2(t) + u_2^2(t) \right] dt + p||x_p(t_f) - X||,$$

where $p > 0$, $x_p = (x, y)$ and $X \in \mathbb{R}^2$ is a terminal position, while satisfying the following constraints:

Constraint 1 (Safety constraint): The robot should avoid collision with a circular obstacle (see Fig. 3.2) defined by the safety constraint (3.12).

Constraint 2 (Robot Limitations): The state and control limitations are defined as: $v_{min} \leq v \leq v_{max}$, $\phi_{min} \leq \phi \leq \phi_{max}$, $u_{1,min} \leq u_1 \leq u_{1,max}$, $u_{2,min} \leq u_2 \leq u_{2,max}$,   where $v_{min} \geq 0$, $v_{max} > 0$, $\phi_{min} \geq 0$, $\phi_{max} > 0$, $u_{1,min} < 0$, $u_{1,max} > 0$, and $u_{2,min} < 0$, $u_{2,max} > 0$.

The relative degrees of (3.12) with respect to (3.11) are 2 and 3 corresponding to u_2 and u_1, respectively, and the relative degree set $S = \{3, 2\}$. We use iHOCBFs to implement (3.12), and use a CLF to enforce the desired terminal state in the objective.

We conducted simulations in MATLAB to compare the effectiveness and performance of the proposed HOCBFs for systems with multiple control inputs. The simulation parameters are $v_d = 5m/s$ (desired speed), $\Delta t = 0.1s$, $(x_d, y_d) = (65, 15)$, $r = 5m$, $M = 1650kg$, $\phi_{max} = -\phi_{min} = 0.6981 rad/s$, $v_{max} = 5m/s$, $v_{min} = 0$, $u_{1,max} = -u_{1,min} = 0.3491$ rad/s^2, $u_{2,max} = -u_{2,min} = 3M$. The robot initial state vector is $x(t_0) = (5m, 15m, 0, 0, 0)$.

We also consider the case of implementing the safety constraint (3.12) with a standard HOCBF ($m = 2$, i.e., using (3.13)) to make a comparison between the original HOCBF and the iHOCBFs. The simulation trajectories for all cases are shown in Fig. 3.3a–b.

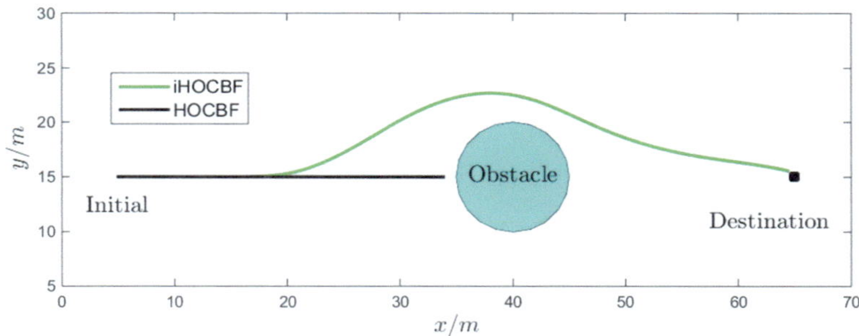

Fig. 3.2 Robot trajectories under different forms of HOCBFs

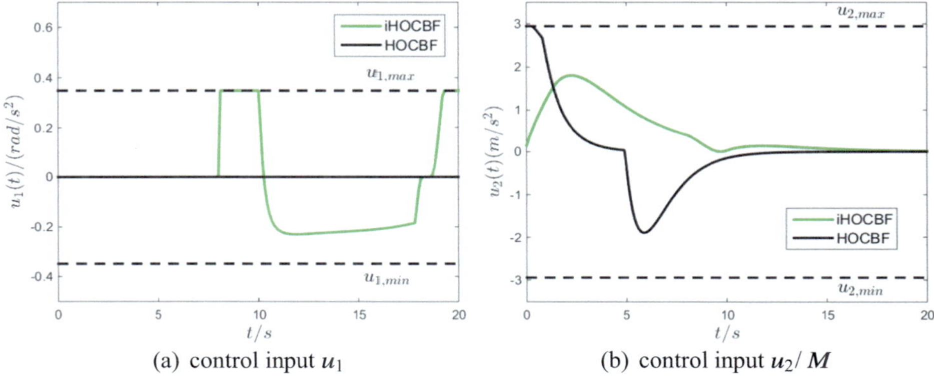

(a) control input u_1 (b) control input u_2/M

Fig. 3.3 Angular and linear accelerations under different forms of HOCBFs

As shown in Fig. 3.2, if we implement the safety constraint (3.12) with a standard HOCBF, only the control input u_2 shows up in the HOCBF constraint (3.13). In other words, the robot can only use deceleration to avoid the obstacle, and thus it cannot get to the destination. This is also demonstrated in its control profile in Fig. 3.3a with $u_1(t) = 0, \forall t \in [t_0, t_f]$ (black solid line).

Feasibility Guarantees for CBF-Based Optimal Control

4

4.1 The Penalty Method

Recall that the construction of a HOCBF in Definition 3.3 relies on the sequence of the $\psi_i(x)$ functions defined through (3.3). The penalty method is based on multiplying the class $\mathcal{K}$ functions $\alpha_i(\cdot)$ in Eq. (3.3) with penalties (weights) $p_i \geq 0, i \in \{1, \ldots, m\}$ to obtain:

$$\psi_i(x) = \dot{\psi}_{i-1}(x) + p_i \alpha_i(\psi_{i-1}(x)), \quad i \in \{1, \ldots, m\}. \tag{4.1}$$

Since we assumed that the control set U and the state space X are closed, we can define

$$U_{\min} := \inf_{x \in X, u \in U} [-L_g L_f^{m-1} b(x(t))u],$$

$$U_{\max} := \sup_{x \in X, u \in U} [-L_g L_f^{m-1} b(x(t))u],$$

$$F_{min} := \inf_{x \in X} [L_f^m b(x)].$$

The following theorem provides conditions for guaranteeing, through the use of penalties, the feasibility of each QP (2.21) used to solve the optimization problem **OCP-CBF** (2.20):

Theorem 4.1 ([72]) *If $U_{\max} \leq F_{min}$, then there exist (small enough) $p_i \geq 0, i \in \{1, \ldots, m\}$ such that the control limitations (2.19) do not conflict with the HOCBF constraint (3.8), $\forall x(t_0) \in C_1 \cap \cdots \cap C_m$.*

© The Author(s), under exclusive license to Springer Nature Switzerland AG 2023
W. Xiao et al., *Safe Autonomy with Control Barrier Functions*,
Synthesis Lectures on Computer Science,
https://doi.org/10.1007/978-3-031-27576-0_4

Proof It follows from the sequence of equations in (4.1) that $p_1, p_2, \ldots, p_{m-1}$ will appear in all terms of $O(b(\boldsymbol{x}))$ in (3.8), i.e., $O(b(\boldsymbol{x})) = 0$ if $p_i = 0, \forall i \in 1, 2, \ldots, m - 1$. Since p_m shows up in the last equation of (4.1), we have

$$-L_g L_f^{m-1} b(\boldsymbol{x}) \boldsymbol{u} \leq L_f^m b(\boldsymbol{x})$$

if $p_i = 0, \forall i \in 1, 2, \ldots, m$. Since $L_f^m b(\boldsymbol{x}) \geq F_{min}$, if $-L_g L_f^{m-1} b(\boldsymbol{x}) \boldsymbol{u} \leq F_{min}$, then the last constraint is satisfied.

The control bound on $\boldsymbol{u}$ in (2.19) always satisfies

$$U_{min} \leq -L_g L_f^{m-1} b(\boldsymbol{x}) \boldsymbol{u} \leq U_{max}.$$

Since $F_{min} \geq U_{max}$, the intersection of the sets determined by the last two inequalities is always non-empty, i.e., the intersection of the control bounds (2.19) and the HOCBF constraint (3.8) is always non-empty. We conclude that there exist small enough penalties $p_1 \geq 0, p_2 \geq 0, \ldots, p_m \geq 0$ such that the control limitations (2.19) will not conflict with the HOCBF constraint (3.8). $\square$

The following corollary provides simpler conditions for some systems (e.g., (2.8) in Example 2.1) that satisfy extra properties:

> **Corollary 4.1** ([72]) *If $\boldsymbol{0} \in U$ and $L_f^m b(\boldsymbol{x}) \geq 0, \forall \boldsymbol{x} \in X$, then there exist (small enough) $p_i \geq 0, i \in \{1, \ldots, m\}$ such that the control limitations (2.19) do not conflict with the HOCBF constraint (3.8), $\forall \boldsymbol{x}(t_0) \in C_1 \cap \cdots \cap C_m$.*

Proof Similar to the proof of the last theorem, we have

$$-L_g L_f^{m-1} b(\boldsymbol{x}) \boldsymbol{u} \leq L_f^m b(\boldsymbol{x}),$$

if $p_i = 0, \forall i \in 1, 2, \ldots, m$. Since $L_f^m b(\boldsymbol{x}) \geq 0, \forall \boldsymbol{x} \in X$, if $-L_g L_f^{m-1} b(\boldsymbol{x}) \boldsymbol{u} \leq 0$, then the last constraint is satisfied. The $\boldsymbol{0}$ vector is included in the last equation, and $\boldsymbol{0} \in U$. Therefore, there exist small enough penalties $p_1 \geq 0, p_2 \geq 0, \ldots, p_m \geq 0$ such that the control limitations (2.19) do not conflict with the HOCBF constraint (3.8). $\square$

Example 4.1 (*Example 2.1 revisited*) For the ACC problem introduced in Example 2.1, $L_f^2 b(\boldsymbol{x}) = 0$. If 0 is included in the control bound, then, from Corollary 4.1, it follows that the HOCBF constraints do not conflict with the control bound when we choose small enough penalties p_1, p_2 for $\alpha_1(\cdot), \alpha_2(\cdot)$. ∎

In order to apply the penalty method in practice, suppose that for some given class $\mathcal{K}$ functions $\alpha_i(\cdot)$, $i \in \{1, \ldots, m\}$ in a HOCBF $b(\boldsymbol{x})$, we observe that the QP (2.21) becomes infeasible for some $\boldsymbol{x} \in C_1 \cap \cdots \cap C_m$, or it becomes infeasible at some time $t \in [t_0, t_f]$. Then, we restart from time t_0 and add penalties to the class $\mathcal{K}$ functions as in (4.1). Note that, when penalties are applied, the sets C_i, $i = 2, \ldots, m$ will be affected. By random selection, we try to find penalty values such that $\boldsymbol{x}(t_0) \in C_1 \cap \cdots \cap C_m$. If the optimization problem becomes feasible, then we are done. Otherwise, we decrease the value of p_1, as p_1 shows up in all the $\psi_i(\cdot)$ functions in (4.1), hence decreasing p_1 is the most efficient way among all the penalties to make the problem feasible. However, decreasing p_1 can significantly shrink C_2, which might result in $\boldsymbol{x}(t_0) \notin C_2$ as shown in (3.4). In order to avoid this, we can proceed by decreasing p_2, and recursively try to find penalties such that $\boldsymbol{x}(t_0) \in C_1 \cap \cdots \cap C_m$.

Example 4.2 (*Example* 3.1 *revisited*) We return to the ACC problem in Example 3.1. The problem setup and model parameters are the same as in Sect. 3.2. We define $b(\boldsymbol{x}) = z - \delta$ as a HOCBF with $m = 2$. Since $L_f^2 b(\boldsymbol{x}) = 0$, the conditions in Corollary 4.1 are satisfied and we can find small enough p_1, p_2 such that the problem is feasible. The control input in this case for quadratic class $\mathcal{K}$ functions is shown in Fig. 4.1. The dashed lines denote the values of the right-hand side of the HOCBF constraint (3.8), i.e., rewriting (3.8) as

$$u \leq \frac{L_f^m b(\boldsymbol{x}) + O(b(\boldsymbol{x})) + \alpha_m(\psi_{m-1}(\boldsymbol{x}))}{-L_g L_f^{m-1} b(\boldsymbol{x})}$$

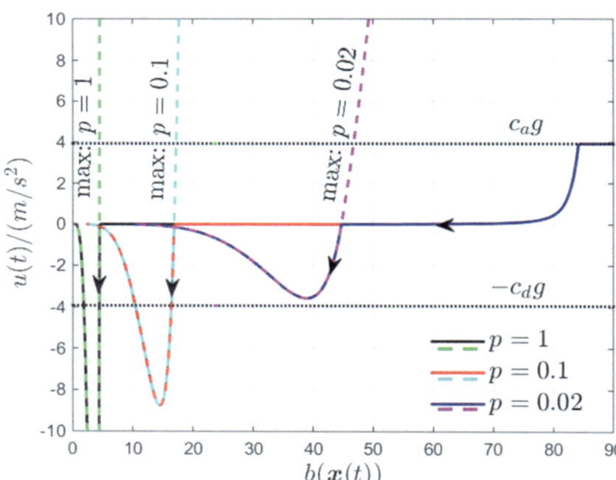

Fig. 4.1 Control input $u(t)$ as $b(\boldsymbol{x}(t)) \to 0$ for different p when using a quadratic class $\mathcal{K}$ function. All the solid lines (black, red, blue) start from $b(\boldsymbol{x}) = 90$. They coincide before the corresponding HOCBF constraint becomes active (e.g., the red solid line can only be seen after $b(\boldsymbol{x}) <= 45$), when the solid line starts overlapping with its associated dashed line. The arrows denote the changing trend for $b(\boldsymbol{x}(t))$ with respect to time

and the solid lines are the optimal controls. When the dashed lines and solid lines coincide, the HOCBF constraint for $b(x)$ is active. In Fig. 4.1, the HOCBF constraint does not conflict with the braking limitation $-c_d g$ when $p = 0.02$ for a quadratic class $\mathcal{K}$ function. The minimum control input (negative) increases as p decreases.   ∎

4.2   Sufficient Conditions for Feasibility

Recall that the discretization method introduced at the end of Chap. 2 for solving the CBF/HOCBF-based optimization problem **OCP-CBF** (2.20) leads to the sequence of QPs (2.21). Consider any such QP over a time step $[\bar{t}, \bar{t} + \Delta t)$. After we apply the constant $u^*(\bar{t})$ to system (2.11) starting at $x(\bar{t})$ for this whole interval $[\bar{t}, \bar{t} + \Delta t)$, it is possible that we end up at a state where the HOCBF constraint (3.8) conflicts with the control bounds (2.19); this clearly would render the QP corresponding to the *next* time interval infeasible.[1] To avoid this, our approach in this section is to define an additional *feasibility constraint* $b_F(x) \geq 0$. Note that this approach is more general than the penalty method in the previous section; however, it usually requires an additional effort to find proper sufficient conditions for feasibility.

> **Definition 4.1** (*Feasibility Constraint*) Suppose the QP **OCP-CBF** (2.20) is feasible at the current state $x(\bar{t}), \bar{t} \in [0, T)$. A constraint $b_F(x) \geq 0$, where $b_F : \mathbb{R}^n \to \mathbb{R}$, is a *feasibility constraint* if it makes the QP corresponding to the next time interval feasible.

In order to ensure that the QP **OCP-CBF** (2.20) is feasible for the next time interval, a feasibility constraint $b_F(x) \geq 0$ should have two important features: (i) it guarantees that the HOCBF constraint (3.8) and the control bound (2.19) do not conflict, (ii) the feasibility constraint $b_F(x) \geq 0$ itself does not conflict with both (3.8) and (2.19) at the same time.

An illustrative example of how a feasibility constraint works is shown in Fig. 4.2. An agent (e.g., robot) whose control is determined by solving a sequence of QPs **OCP-CBF** (2.20) is about to get close to an obstacle at the following step. The next state may be infeasible for the QP associated with that next step. For example, the state denoted by the red circle in Fig. 4.2 may include a large speed component such that the agent cannot find a control (deceleration or heading) to avoid the obstacle at the next step. However, if we include a feasibility constraint that can prevent the agent from reaching this state, then this next QP is feasible.

After we determine a feasibility constraint of the form $b_F(x) \geq 0$, we can enforce it through a CBF and take it as an additional constraint for **OCP-CBF** (2.20) in order to guarantee feasibility given a system state x. We show how we can determine an appropriate feasibility constraint in the sequel. First, we provide a simple example to illustrate the need for a feasibility constraint when dealing with the CBF-CLF-based QPs **OCP-CBF** (2.20).

---

[1] Note that, since the CLF constraint is relaxed, it does not affect the feasibility of the QP.

Fig. 4.2 An illustration of how a feasibility constraint works for an obstacle avoidance problem. Adding an appropriate feasibility constraint at the current state prevents the agent from going into the infeasible next state

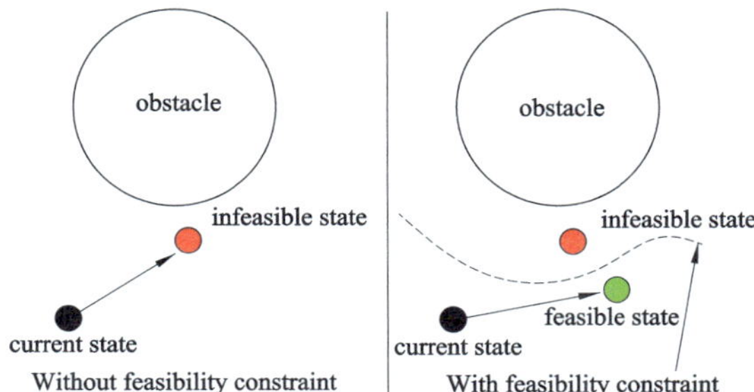

Example 4.3 (*Adaptive Cruise Control*) This example is based on the simple ACC problem in Example 2.1, but considers a more complex model for the vehicle dynamics:

$$\underbrace{\begin{bmatrix} \dot{v}(t) \\ \dot{z}(t) \end{bmatrix}}_{\dot{x}(t)} = \underbrace{\begin{bmatrix} -\frac{1}{M} F_r(v(t)) \\ v_p - v(t) \end{bmatrix}}_{f(x(t))} + \underbrace{\begin{bmatrix} \frac{1}{M} \\ 0 \end{bmatrix}}_{g(x(t))} u(t), \tag{4.2}$$

where M denotes the mass of the ego vehicle, $z(t)$ denotes the distance between the preceding and the ego vehicles, $v_p \geq 0$, $v(t) \geq 0$ denote the speeds of the preceding and the ego vehicles, respectively, and $F_r(v(t))$ denotes the resistance force, which is expressed [26] as:

$$F_r(v(t)) = f_0 sgn(v(t)) + f_1 v(t) + f_2 v^2(t),$$

where $f_0 > 0$, $f_1 > 0$ and $f_2 > 0$ are scalars determined empirically. The first term in $F_r(v(t))$ denotes the Coulomb friction force, the second term denotes the viscous friction force and the last term denotes the aerodynamic drag. The control $u(t)$ is the driving force of the ego vehicle subject to the constraint:

$$-c_d M g \leq u(t) \leq c_a M g, \quad \forall t \geq 0, \tag{4.3}$$

where $c_a > 0$ and $c_d > 0$ are the maximum acceleration and deceleration coefficients, respectively, and g is the gravity constant.

We require that the distance $z(t)$ between the ego vehicle and its immediately preceding vehicle be greater than $l_0 > 0$, i.e.,

$$z(t) \geq l_0, \quad \forall t \geq 0. \tag{4.4}$$

Let $b(x(t)) := z(t) - l_0$. The relative degree of $b(x(t))$ is $m = 2$, so we choose a HOCBF following Definition 3.3 by defining $\psi_0(x(t)) := b(x(t))$, $\alpha_1(\psi_0(x(t))) := p_1 \psi_0(x(t))$ and $\alpha_2(\psi_1(x(t))) := p_2 \psi_1(x(t))$, $p_1 > 0$, $p_2 > 0$. We then seek a control for the ego vehicle such that the constraint $b(x(t)) \geq 0$ is satisfied. The control $u(t)$ should satisfy (3.8), which in this case is:

$$\underbrace{\frac{F_r(v(t))}{M}}_{L_f^2 b(\boldsymbol{x}(t))} + \underbrace{\frac{-1}{M}}_{L_g L_f b(\boldsymbol{x}(t))} \times u(t) + \underbrace{p_1(v_p - v(t))}_{O(b(\boldsymbol{x}(t)))}$$

$$\underbrace{+ \, p_2(v_p - v(t)) + p_1 p_2(z(t) - l_0)}_{\alpha_2(\psi_1(\boldsymbol{x}(t)))} \geq 0. \tag{4.5}$$

Suppose we wish to minimize $\int_0^T \left(\frac{u(t) - F_r(v(t))}{M} \right)^2 dt$ in the constrained optimal control problem **OCP-CBF** (2.20). Applying the QP-based method **OCP-CBF** (2.20), note that the HOCBF constraint above can easily conflict with $-c_d Mg \leq u(t)$ in (4.3), i.e., the ego vehicle cannot brake in time under control constraint (4.3) so that the safety constraint (4.4) is satisfied when the two vehicles get close to each other. This is intuitive to see when we rewrite (4.5) in the form of an acceleration constraint:

$$\frac{1}{M} u(t) \leq \frac{F_r(v(t))}{M} + (p_1 + p_2)(v_p - v(t)) + p_1 p_2(z(t) - l_0). \tag{4.6}$$

The right-hand side above is usually negative when the two vehicles are close to each other. If it is smaller than $-c_d Mg$, the HOCBF constraint (4.5) will conflict with $-c_d Mg \leq u(t)$ in (4.3). When this happens while solving a specific QP, this QP will be infeasible. ∎

In the rest of this section, we show how we can solve in general terms the infeasibility problem illustrated in the example above by constructing a feasibility constraint as in Definition 4.1.

4.2.1 Feasibility Conditions for Constraints with Relative Degree One

For ease of exposition, we start our analysis with feasibility constraints for a relative-degree-one safety constraint, and then generalize it to the case of high-relative-degree safety constraints. It will become clear, however, from the overall analysis that the determination of an appropriate $b_F(\boldsymbol{x})$ does not depend on the relative degree of the constraints.

Suppose we have a constraint $b(\boldsymbol{x}) \geq 0$ with relative degree one for system (2.11), where $b : \mathbb{R}^n \to \mathbb{R}$. Then, we can define $b(\boldsymbol{x})$ as a HOCBF with $m = 1$ as in Definition 3.3, i.e., we have a "traditional" first order CBF. Following (2.13), any control $\boldsymbol{u} \in U$ should satisfy the CBF constraint:

$$- L_g b(\boldsymbol{x})\boldsymbol{u} \leq L_f b(\boldsymbol{x}) + \alpha(b(\boldsymbol{x})), \tag{4.7}$$

where $\alpha(\cdot)$ is a class $\mathcal{K}$ function of its argument. We define a set of controls that satisfy the last equation as:

$$K(\boldsymbol{x}) = \{\boldsymbol{u} \in \mathbb{R}^q : -L_g b(\boldsymbol{x})\boldsymbol{u} \leq L_f b(\boldsymbol{x}) + \alpha(b(\boldsymbol{x}))\}. \tag{4.8}$$

Our analysis for determining a feasibility constraint depends on whether any component of the vector $L_g b(x)$ will change sign within the time interval $[0, T]$ or not.

1. The case when all components in $L_g b(x)$ do not change sign

In this case, the inequality constraint for each control component does not change sign if we multiply each component of $L_g b(x)$ by the corresponding one of the control bounds in (2.19). Therefore, we assume that $L_g b(x) \leq 0$ (componentwise), $0 \in \mathbb{R}^q$ in the rest of this section. The analysis for other cases (each component of $L_g b(x)$ is either non-negative or non-positive) is similar. Not all the components in $L_g b(x)$ can be 0 due to the relative degree definition in Definition 3.1. We can multiply the control bounds (2.19) by the vector $-L_g b(x)$, and get

$$- L_g b(x) u_{\min} \leq -L_g b(x) u \leq -L_g b(x) u_{\max}. \tag{4.9}$$

The control constraint (4.9) is actually a relaxation of the control bound (2.19) as we multiply each component of $L_g b(x)$ by the corresponding one of the control bounds in (2.19), and then add them together. We define

$$U_{ex}(x) = \{u \in \mathbb{R}^q : -L_g b(x) u_{\min} \leq -L_g b(x) u \leq -L_g b(x) u_{\max}\}. \tag{4.10}$$

We also provide the following formal definition describing how two or more state-dependent control constraints are "conflict-free":

Definition 4.2 (*Conflict-free control constraints*) We say that two (or more) state-dependent control constraints are *conflict-free* if the intersection of the two (or more) sets defined by these constraints in terms of u is non-empty for all $x \in X$.

Looking at (4.10), it is obvious that U is a subset of $U_{ex}(x)$. An example of a two-dimensional control $u = (u_1, u_2)$ is shown in Fig. 4.3. Nonetheless, the relaxation set $U_{ex}(x)$ does not negatively affect the property of the following lemma:

Lemma 4.1 ([79]) *If the control u in (2.11) is such that (4.9) is conflict-free with (4.7) for all $x \in X$, then the control bound (2.19) is also conflict-free with (4.7).*

Proof Let $g = (g_1, \ldots, g_q)$ in (2.11), where $g_i : \mathbb{R}^n \to \mathbb{R}^n, i, \in \{1, \ldots, q\}$ and we can write: $L_g b(x) = (L_{g_1} b(x), \ldots, L_{g_q} b(x)) \in \mathbb{R}^{1 \times q}$. For the control bounds $u_{i,min} \leq u_i \leq u_{i,max}, i \in \{1, \ldots, q\}$ in (2.19), we can multiply by $-L_{g_i} b(x)$ and get

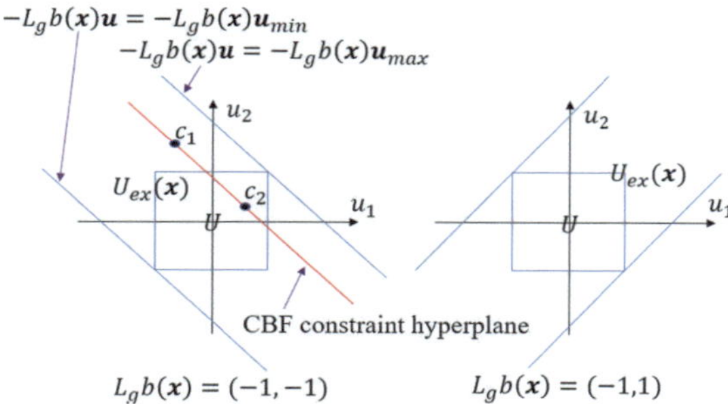

Fig. 4.3 The relationship between $U \subset U_{ex}(x)$ and $U_{ex}(x)$ in the case of a two-dimensional control $u = (u_1, u_2)$. The magnitude of $L_g b(x)$ determines the slopes of the two lines (hyperplanes) $-L_g b(x)u = -L_g b(x)u_{max}$ and $-L_g b(x)u = -L_g b(x)u_{min}$. If there exists a control $c_1 \in U_{ex}(x)$ that satisfies the CBF constraint (4.7) (on the boundary), then there exists a control $c_2 \in U$ that also satisfies the CBF constraint (4.7) (on the boundary)

$$-L_{g_i} b(x)u_{i,min} \leq -L_{g_i} b(x)u_i \leq -L_{g_i} b(x)u_{i,max}, \quad i \in \{1, \ldots, q\},$$

as we have assumed that $L_g b(x) \leq 0$. If we take the summation of the inequality above over all $i \in \{1, \ldots, q\}$, then we obtain the constraint (4.9). Therefore, the satisfaction of (2.19) implies the satisfaction of (4.9). Then, U defined in (2.19) is a subset of $U_{ex}(x)$. It is obvious that the boundaries of the set $U_{ex}(x)$ in (4.10) and $K(x)$ in (4.8) are hyperplanes, and these boundaries are parallel to each other for all $x \in X$. Meanwhile, the two boundaries of $U_{ex}(x)$ pass through the two corners u_{min}, u_{max} of the set U (a polyhedron) following (4.10), respectively. If there exists a control $c_1 \in U_{ex}(x)$ (e.g., in Fig. 4.3) that satisfies (4.7), then the boundary of the set $K(x)$ in (4.8) lies either between the two hyperplanes defined by $U_{ex}(x)$ or above these two hyperplanes (i.e., $U_{ex}(x)$ is a subset of $K(x)$ in (4.8)). In the latter case, the lemma is true as U is a subset of $U_{ex}(x)$. In the former case, we can always find another control $c_2 \in U$ (e.g., in Fig. 4.3) that satisfies (4.7) as the boundary of $K(x)$ in (4.8) is parallel to the two $U_{ex}(x)$ boundaries that respectively pass through the two corners u_{min}, u_{max} of the set U. Therefore, although U is a subset of $U_{ex}(x)$, it follows that if (4.9) is conflict-free with (4.7) in terms of u for all $x \in X$, the control bound (2.19) is also conflict-free with (4.7). $\qquad\square$

Motivated by Lemma 4.1, in order to determine if (4.7) complies with (2.19), we may just consider (4.7) and (4.9). Since there are two inequalites in (4.9), we have two cases to consider: $(i) - L_g b(x)u \leq -L_g b(x)u_{max}$ and (4.7) and $(ii) - L_g b(x)u_{min} \leq -L_g b(x)u$ and (4.7). It is obvious that there always exists a control u such that the two inequalities in case (i) are satisfied for all $x \in X$, while this may not be true for case (ii), depending on x. For example, the CBF for the rear-end safety constraint (2.5) in the ACC problem of Example 2.1 may conflict with the maximum braking force $-c_d M g < 0$, and it will never conflict with the maximum driving force $c_a M g > 0$ as the ego vehicle needs to brake when it gets close to the preceding vehicle in order to satisfy the safety constraint (2.5). Therefore, in terms of avoiding the conflict between the CBF constraint (4.7) and control bounds (4.9) that leads to the infeasibility of problem **OCP-CBF** (2.20), we wish to satisfy:

$$L_f b(x) + \alpha(b(x)) \geq -L_g b(x) u_{\min}. \tag{4.11}$$

This is called the **feasibility constraint** for problem **OCP-CBF** (2.20) in the case of a relative-degree-one safety constraint $b(x) \geq 0$ in (2.18). Observe that the relative degree of the feasibility constraint (4.11) is also one with respect to dynamics (2.11), since it is entirely dependent on $b(x)$.

The next step is to determine a control such that the feasibility constraint (4.11) is guaranteed to be satisfied. Towards this goal, we treat (4.11) as a constraint to be satisfied and define an associated candidate CBF $b_F(x)$ just like in Definition 3.3:

$$b_F(x) = L_f b(x) + \alpha(b(x)) + L_g b(x) u_{\min} \geq 0. \tag{4.12}$$

Then, if $b_F(x(0)) \geq 0$, we can obtain a feedback controller $u \in K_F(x)$ defined below that guarantees the CBF constraint (4.7) and the control bounds (2.19) do not conflict with each other:

$$K_F(x) = \{u \in \mathbb{R}^q : L_f b_F(x) + L_g b_F(x) u + \alpha_f(b_F(x)) \geq 0\}, \tag{4.13}$$

where $\alpha_f(\cdot)$ is a class $\mathcal{K}$ function. This is formalized in the following theorem.

> **Theorem 4.2** ([79]) *If **OCP** is initially feasible and the CBF constraint in (4.13) corresponding to (4.11) does not conflict with both the control bounds (2.19) and (4.7) at the same time, any controller $u \in K_F(x)$ guarantees the feasibility of problem **OCP-CBF** (2.20).*

Proof If **OCP** is initially feasible, then the CBF constraint (4.7) for the safety requirement (2.18) does not conflict with the control bounds (2.19) at time 0. It also does not conflict with the constraint (4.9) as U is a subset of $U_{ex}(x)$ that is defined in (4.10). In other words, $b_F(x(0)) \geq 0$ holds in the feasibility constraint (4.11). Thus, the initial condition for the CBF in Definition 3.3 is satisfied. By Theorem 3.2, we have that $b_F(x(t)) \geq 0, \forall t \geq 0$. Therefore, the CBF constraint (4.7) does not conflict with the constraint (4.9) for all $t \geq 0$. By Lemma 4.1, the CBF constraint (4.7) also does not conflict with the control bound (2.19). Finally, since the CBF constraint in (4.13) corresponding to (4.11) does not conflict with the control bounds (2.19) and (4.7) at the same time by assumption, we conclude that the feasibility of the problem is guaranteed.   □

The condition that the CBF constraint in (4.13) corresponding to (4.11) does not conflict with both the control bounds (2.19) and (4.7) at the same time from Theorem 4.2 is strong. Moreover, even if this condition is not satisfied, then the problem can still be infeasible. In order to relax this condition, one option is to recursively define other new feasibility constraints for the feasibility constraint (4.11) to address the possible conflict between (4.13) and (2.19), and (4.7). However, the number of iterations is not bounded, thus leading to a potentially large (unbounded) set of feasibility constraints.

In order to address the unbounded iteration issue in finding feasibility constraints, we will instead try to express the feasibility constraint in (4.13) so that it is in a form that is similar to that of the CBF constraint (4.7). If this is achieved, we can make these two constraints compliant with each other, thus addressing the unbounded iteration issue mentioned above. Therefore, we try to construct the CBF constraint in (4.13) so that it takes the form:

$$L_f b(\boldsymbol{x}) + L_g b(\boldsymbol{x})\boldsymbol{u} + \alpha(b(\boldsymbol{x})) + \varphi(\boldsymbol{x}, \boldsymbol{u}) \geq 0 \tag{4.14}$$

for some appropriately selected function $\varphi(\boldsymbol{x}, \boldsymbol{u})$. One obvious choice for $\varphi(\boldsymbol{x}, \boldsymbol{u})$ immediately following (4.13) is $\varphi(\boldsymbol{x}, \boldsymbol{u}) = L_f b_F(\boldsymbol{x}) + L_g b_F(\boldsymbol{x})\boldsymbol{u} + \alpha_f(b_F(\boldsymbol{x})) - L_f b(\boldsymbol{x}) - L_g b(\boldsymbol{x})\boldsymbol{u} - \alpha(b(\boldsymbol{x}))$, which can be simplified through a proper choice of the class $\mathcal{K}$ functions $\alpha(\cdot), \alpha_f(\cdot)$, as will be shown next. Since we will eventually include the constraint $\varphi(\boldsymbol{x}, \boldsymbol{u}) \geq 0$ into our QPs (shown later) to address the infeasibility problem, we wish its relative degree to be low. Otherwise, it becomes necessary to use HOCBFs to make the control show up in enforcing $\varphi(\boldsymbol{x}) \geq 0$ (instead of $\varphi(\boldsymbol{x}, \boldsymbol{u}) \geq 0$ due to its high relative degree), which could make the corresponding HOCBF constraint complicated, and make it easily conflict with the control bound (2.19) and the CBF constraint (4.7), hence leading to the infeasibility of the QPs. Therefore, we define a candidate function as follows (note that a relative-degree-zero function means that the control $\boldsymbol{u}$ directly shows up in the function itself):

> **Definition 4.3** (*Candidate φ function*) A function $\varphi(\boldsymbol{x}, \boldsymbol{u})$ in (4.14) is a *candidate φ function* if its relative degree with respect to the systems dynamics (2.11) is either one or zero.

Our task, therefore, is to determine a candidate φ function. Based on the reformulation of the CBF constraint in (4.13), we can properly choose the class $\mathcal{K}$ function $\alpha(\cdot)$ in (4.7). A typical choice for $\alpha(\cdot)$ is a linear function, in which case we automatically have the constraint formulation (4.14) by substituting the function $b_F(\boldsymbol{x})$ from (4.12) into (4.13), and get

$$\varphi(\boldsymbol{x}, \boldsymbol{u}) = L_f^2 b(\boldsymbol{x}) + L_g L_f b(\boldsymbol{x})\boldsymbol{u} + L_f(L_g b(\boldsymbol{x})\boldsymbol{u}_{min})$$
$$+ L_g(L_g b(\boldsymbol{x})\boldsymbol{u}_{min})\boldsymbol{u} + \alpha_f(b_F(\boldsymbol{x})) - b(\boldsymbol{x}).$$

Note that it is possible that $L_g L_f b(x) = 0$ and $L_g(L_g b(x) u_{min}) = 0$, depending on the dynamics (2.11) and the CBF $b(x)$. In this case the relative degree of $\varphi(x, u)$ (written as $\varphi(x)$) is one, since $\alpha_f(b_F(x))$ is contained in it and $b_F(x)$ is a function of $b(x)$.

If the relative degree of $\varphi(x, u)$ is zero (e.g., $L_g L_f b(x) = 0$ and $L_g(L_g b(x) u_{min}) = 0$ are not satisfied above), we require that

$$\varphi(x, u) \geq 0, \tag{4.15}$$

such that the satisfaction of the CBF constraint (4.7) implies the satisfaction of the CBF constraint (4.14), and the satisfaction of the CBF constraint (4.14) implies the satisfaction of (4.11) by Theorem 3.2, i.e., the CBF constraint (4.7) does not conflict with the control bound (2.19). Moreover, if (4.15) happens to not conflict with both (4.7) and (2.19) at the same time, depending on the CBF $b(x)$ and the dynamics (2.11), then the QPs are guaranteed to be feasible.

The CBF constraint (4.13) for the feasibility constraint is similar to the CBF constraint (4.7) for safety by properly defining the class $\mathcal{K}$ functions α, α_f, which generates (4.15) that needs to be satisfied. Therefore, the QP feasibility can be improved, and even be guaranteed, if constraint (4.15) satisfies similar conditions as shown in Theorem 4.3. This is more helpful in the case of safety constraints with high relative degree (considered in the next section) as the HOCBF constraint (3.8) has many complicated terms; therefore, it is better to remove these terms in the feasibility constraint and just consider (4.15) in the QP in order to make (4.15) compliant with (4.7) and (2.19).

If the relative degree of a candidate $\varphi(x, u)$ with respect to (2.11) is one, i.e., $\varphi(x, u) \equiv \varphi(x)$, we define a set $U_s(x)$:

$$U_s(x) = \{u \in \mathbb{R}^q : L_f \varphi(x) + L_g \varphi(x) u + \alpha_u(\varphi(x)) \geq 0\} \tag{4.16}$$

where $\alpha_u(\cdot)$ is a class $\mathcal{K}$ function. Within the set of candidate functions $\varphi(x)$, if we can find one that satisfies the conditions of the following theorem, then the feasibility of problem **OCP-CBF** (2.20) is guaranteed:

Theorem 4.3 ([79]) *If $\varphi(x)$ is a candidate φ function such that $\varphi(x(0)) \geq 0$, $L_f \varphi(x) \geq 0$, $L_g \varphi(x) = \gamma L_g b(x)$, for some $\gamma > 0$, $\forall x \in X$ and $\mathbf{0} \in U$, then any controller $u(t) \in U_s(x)$, $\forall t \geq 0$ guarantees the feasibility of problem **OCP-CBF** (2.20).*

Proof Since $\varphi(x)$ is a candidate φ function, we can define a set $U_s(x)$ as in (4.16). If $\varphi(x(0)) \geq 0$ and $u(t) \in U_s(x)$, $\forall t \geq 0$, we have that $\varphi(x(t)) \geq 0$, $\forall t \geq 0$ by Theorem 3.2. Then, the satisfaction of the CBF constraint (4.7) corresponding to the safety constraint (2.18) implies the satisfaction of the CBF constraint (4.14) (equivalent to (4.13)) for the feasibility constraint (4.11). In other words, the CBF constraint (4.7) automatically guarantees that it will not conflict with the control constraint (4.9) as the satisfaction of (4.14) implies the satisfaction of (4.11) following Theorem 3.2 and (4.11) guarantees that (4.7) and (4.9) are conflict-free. By Lemma 4.1, the CBF constraint (4.7) will also not conflict with the control bound U in (2.19), i.e. $K(x) \cap U \neq \emptyset$, where $K(x)$ is defined in (4.8).

Since $L_f \varphi(x) \geq 0$, we have that $\mathbf{0} \in U_s(x)$. We also have $\mathbf{0} \in U(x)$, thus, $U_s(x) \cap U \neq \emptyset$ is guaranteed. Since $L_g \varphi(x) = \gamma L_g b(x)$, $\gamma > 0$, the two hyperplanes of the two half spaces formed by $U_s(x)$ in (4.16) and $K(x)$ in (4.8) are parallel to each other, and the normal directions of the two hyperplanes along the half space direction are the same. Thus, $U_s(x) \cap K(x)$ is either $U_s(x)$ or $K(x)$, i.e., either $U_s(x) \cap K(x) \cap U = U_s(x) \cap U$ or $U_s(x) \cap K(x) \cap U = K(x) \cap U$. As $U_s(x) \cap U \neq \emptyset$ and $K(x) \cap U \neq \emptyset$, we have $U_s(x) \cap K(x) \cap U \neq \emptyset, \forall x \in X$. Therefore, the CBF constraint (4.7) does not conflict with the control bound (2.19) and the CBF constraint in $U_s(x)$ at the same time, and we can conclude that the problem is guaranteed to be feasible. $\qquad\square$

The conditions in Theorem 4.3 are **sufficient conditions** for the feasibility of problem **OCP-CBF** (2.20). Under these conditions, we can claim that $\varphi(x) \geq 0$ is a single **feasibility constraint** that guarantees the feasibility of problem **OCP-CBF** (2.20) in the case that the safety constraint (2.18) is with relative degree one, i.e., $m = 1$ in (3.8).

The conditions that the function $\varphi(x)$ must satisfy in Theorem 4.3 may be conservative and determining such a function is a nontrivial problem. For a general system (2.11) and safety constraint (2.18), we can parameterize the definition of the CBF (4.7) for the safety and the CBF constraint for the feasibility constraint (4.13), i.e., parameterize $\alpha(\cdot)$ and $\alpha_F(\cdot)$, such as the form in [74]. We can then choose the parameters to satisfy the conditions in Theorem 4.3.

On the other hand, for some special classes of dynamics (2.11), we can design a systematic way to derive such $\varphi(x)$ functions. In the case of such dynamics, we may even relax some of the conditions in Theorem 4.3. For example, if both the dynamics (2.11) and the safety constraint (2.18) are in linear forms, then the condition $L_g \varphi(x) = \gamma L_g b(x)$, for some $\gamma > 0$ in Theorem 4.3 is satisfied, and thus this condition is removed. An example of determining a function $\varphi(x)$ for the ACC problem in Example 4.3 is given in Example 4.4 at the end of this section.

Returning to the optimization problem **OCP-CBF** (2.20) in which $C(\cdot)$ is quadratic, we can now obtain a feasible version of this problem in the form:

$$\min_{u(t),\delta_r(t)} \int_0^T [\|u(t)\|^2 + p_t \delta_r^2(t)]dt \tag{4.17}$$

subject to the feasibility constraint (4.15) if the relative degree of $\varphi(x, u)$ is 0; otherwise, subject to the CBF constraint in (4.16), and all constraints in **OCP-CBF** (2.20). Here, $\varphi(x)$ satisfies the conditions in Theorem 4.3 for (4.16), and (4.15) is assumed to be non-conflicting with the CBF constraint (4.7) and the control bound (2.19) at the same time. In order to guarantee feasibility, we may try to find a $\varphi(x)$ that has relative degree one and

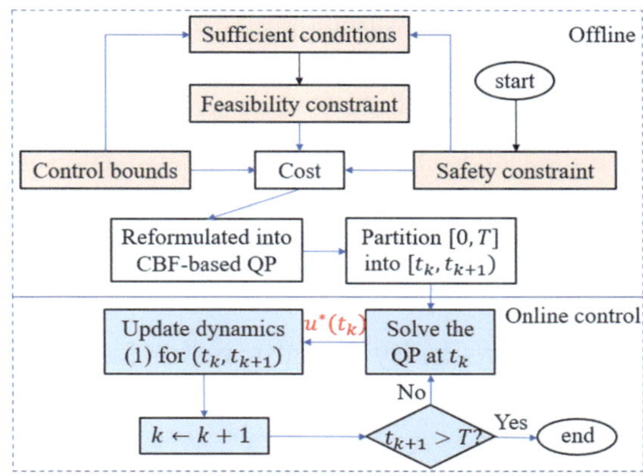

Fig. 4.4 Overall process of solving the constrained optimal control problem with the feasibility guaranteed CBF method

that satisfies the conditions in Theorem 4.3. The overall process of solving the constrained optimal control problem with the feasibility guaranteed CBF method is described in Fig. 4.4.

2. The case when some components in $L_g b(x)$ change sign

Recall that $L_g b(x) = (L_{g_1} b(x), \ldots, L_{g_q} b(x)) \in \mathbb{R}^{1 \times q}$ and let $u = (u_1, \ldots, u_q)$, $u_{\min} = (u_{1,\min}, \ldots, u_{q,\min}) \leq 0$, $u_{\max} = (u_{1,\max}, \ldots, u_{q,\max}) \geq 0, 0 \in \mathbb{R}^q$. If $L_{g_i} b(x)$, $i \in \{1, \ldots, q\}$ changes sign in $[0, T]$, then we have the following two (symmetric and non-symmetric) cases to consider in order to find a valid feasibility constraint.

Case 1: the control bound for u_i, $i \in \{1, \ldots, q\}$ is symmetric, i.e. $u_{i,\max} = -u_{i,\min}$. In this case, by multiplying $-L_{g_i} b(x)$ by the control bound for u_i, we have

$$- L_{g_i} b(x) u_{i,\min} \leq -L_{g_i} b(x) u_i \leq -L_{g_i} b(x) u_{i,\max} \tag{4.18}$$

if $L_{g_i} b(x) < 0$. When $L_{g_i} b(x)$ changes sign at some time $t_1 \in [0, T]$, then the sign of the last equation is reversed. However, since $u_{i,\max} = -u_{i,\min}$, we have exactly the same constraint as (4.18), and $-L_{g_i} b(x) u_{i,\min}$ will still be continuously differentiable when we construct the feasibility constraint as in (4.11). Therefore, the feasibility constraint (4.11) will not be affected by the sign change of $L_{g_i} b(x)$, $i \in \{1, \ldots, q\}$.

Case 2: the control bound for u_i, $i \in \{1, \ldots, q\}$ is not symmetric, i.e., $u_{i,\max} \neq -u_{i,\min}$. In this case, we can define:

$$u_{i,\lim} := \min\{|u_{i,\min}|, u_{i,\max}\} \tag{4.19}$$

Considering (4.19), we have the following constraint:

$$- u_{i,\lim} \leq u_i \leq u_{i,\lim}. \tag{4.20}$$

The satisfaction of the last equation implies the satisfaction of $u_{i,\min} \leq u_i \leq u_{i,\max}$ in (2.19).

If $L_{g_i} b(x) < 0$, we multiply the control bound by $-L_{g_i} b(x)$ for u_i and have the following constraint:

$$L_{g_i} b(x) u_{i,\text{lim}} \leq -L_{g_i} b(x) u_i \leq -L_{g_i} b(x) u_{i,\text{lim}} \tag{4.21}$$

The satisfaction of (4.21) implies the satisfaction of (4.18) following (4.19). Now, the control bound for u_i is converted to the symmetric case, and the feasibility constraint (4.11) will not be affected by the sign change of $L_{g_i} b(x)$, $i \in \{1, \ldots, q\}$.

4.2.2 Feasibility Conditions for Safety Constraints with High Relative Degree

Suppose we have a constraint $b(x) \geq 0$ with relative degree $m \geq 1$ for system (2.11), where $b : \mathbb{R}^n \to \mathbb{R}$. Then we can define $b(x)$ as a HOCBF as in Definition 3.3. Any control $u \in U$ should satisfy the HOCBF constraint (3.8). In what follows, we also assume that $L_g L_f^{m-1} b(x) \leq 0$, where $0 \in \mathbb{R}^q$ and all components in $L_g L_f^{m-1} b(x)$ do not change sign in $[0, T]$. The analysis for all other cases is similar to the last section.

Similar to (4.7), we rewrite the HOCBF constraint (3.8) as

$$-L_g L_f^{m-1} b(x) u \leq L_f^m b(x) + O(b(x)) + \alpha_m(\psi_{m-1}(x)) \tag{4.22}$$

and we can multiply the control bounds (2.19) by the vector $-L_g L_f^{m-1} b(x)$:

$$-L_g L_f^{m-1} b(x) u_{\min} \leq -L_g L_f^{m-1} b(x) u \leq -L_g L_f^{m-1} b(x) u_{\max}. \tag{4.23}$$

As in (4.9), the last equation is also a relaxation of the original control bound (2.19), and Lemma 4.1 still applies in the high-relative-degree-constraint case.

The HOCBF constraint (4.22) may conflict with the left inequality of the transformed control bound (4.23) when its right hand side is smaller than $-L_g L_f^{m-1} b(x) u_{\min}$. Therefore, we wish to satisfy

$$L_f^m b(x) + O(b(x)) + \alpha_m(\psi_{m-1}(x)) \geq -L_g L_f^{m-1} b(x) u_{\min}. \tag{4.24}$$

This is called the **feasibility constraint** for the problem **OCP-CBF** (2.20) in the case of a high-relative-degree constraint $b(x) \geq 0$ in (2.18).

In order to find a control such that the feasibility constraint (4.11) is guaranteed to be satisfied, we define

$$b_{hF}(x) = L_f^m b(x) + O(b(x)) + \alpha_m(\psi_{m-1}(x)) + L_g L_f^{m-1} b(x) u_{\min} \geq 0,$$

and define $b_{hF}(x)$ to be a HOCBF as in Definition 3.3. It is important to note that the relative degree of $b_{hF}(x)$ with respect to dynamics (2.11) is only one, as we have $\psi_{m-1}(x)$ in it.

Thus, we can get a feedback controller $K_{hF}(x)$ that guarantees free conflict between the HOCBF constraint (4.22) and the control bounds (2.19):

$$K_{hF}(x) = \{u \in \mathbb{R}^q : L_f b_{hF}(x) + L_g b_{hF}(x)u + \alpha_f(b_{hF}(x)) \geq 0\}, \tag{4.25}$$

if $b_{hF}(x(0)) \geq 0$, where $\alpha_f(\cdot)$ is a class $\mathcal{K}$ function.

Theorem 4.4 ([79]) *If **OCP** is initially feasible and the CBF constraint in (4.25) corresponding to (4.24) does not conflict with control bounds (2.19) and (4.22) at the same time, any controller $u \in K_{hf}(x)$ guarantees the feasibility of problem **OCP-CBF** (2.20).*

Proof The proof is the same as that of Theorem 4.2. □

Similar to the motivation for the analysis of the relative degree one case, we also reformulate the constraint in (4.25) in the form:

$$L_f^m b(x) + L_g L_f^{m-1} b(x)u + O(b(x)) + \alpha_m(\psi_{m-1}(x)) + \varphi(x, u) \geq 0 \tag{4.26}$$

for some appropriate $\varphi(x, u)$. An obvious choice is $\varphi(x, u) = L_f b_{hF}(x) + L_g b_{hF}(x)u + \alpha_f(b_{hF}(x)) - L_f^m b(x) - L_g L_f^{m-1} b(x)u - O(b(x)) - \alpha_m(\psi_{m-1}(x))$, which is a candidate φ function we would like to simplify. We define a set $U_s(x)$ similar to (4.16).

Similar to the last subsection, we only consider the case when the relative degree of $\varphi(x, u)$ is one, i.e., we have $\varphi(x)$ from now on. Then, we have the following theorem to guarantee the feasibility of the problem **OCP-CBF** (2.20):

Theorem 4.5 ([79]) *If $\varphi(x)$ is a candidate function, $\varphi(x(0)) \geq 0, L_f\varphi(x) \geq 0, L_g\varphi(x) = \gamma L_g L_f^{m-1} b(x)$, for some $\gamma > 0, \forall x \in X$ and $0 \in U$, then any controller $u(t) \in U_s(x), \forall t \geq 0$ guarantees the feasibility of the problem **OCP-CBF** (2.20).*

Proof The proof is similar to the one for Theorem 4.3. □

The approach to finding a candidate function $\varphi(x)$ is the same as in the last section. The conditions in Theorem 4.5 are **sufficient conditions** for the feasibility of the problem **OCP-CBF** (2.20). Under the conditions in Theorem 4.5, we can also claim that $\varphi(x) \geq 0$ is a single **feasibility constraint** that guarantees the feasibility of the problem **OCP-CBF** (2.20) in the case that the safety constraint (2.18) has a high relative degree. We can get a feasible problem from the original problem **OCP-CBF** (2.20) (in which $C(\cdot)$ is quadratic) in the form:

$$\min_{u(t),\delta_r(t)} \int_0^T [||u(t)||^2 + p_t \delta_r^2(t)]dt, \tag{4.27}$$

subject to the feasibility constraint (4.15) if the relative degree of $\varphi(x, u)$ is 0; otherwise, subject to the CBF constraint in (4.16), and all constraints in **OCP-CBF** (2.20). Here, the CBF constraint in **OCP-CBF** (2.20) is replaced by the HOCBF constraint (3.8), and $\varphi(x)$ satisfies the conditions in Theorem 4.5 for (4.16), and (4.15) is assumed to be non-conflicting with the HOCBF constraint (3.8) and the control bound (2.19) at the same time.

Remark 4.1 When we have multiple safety constraints, we can employ similar ideas to find sufficient conditions to guarantee problem feasibility. However, we also need to make sure that these sufficient conditions do not conflict with each other.

Example 4.4 (*Example* 4.3 *revisited*) Here we demonstrate how we can find a single feasibility constraint $\varphi(x(t)) \geq 0$ for the ACC problem. Recall that in this problem we have $b(x(t)) = z(t) - l_0$. First, it is obvious that $L_g L_f b(x(t)) = -\frac{1}{M}$ in (4.2) does not change sign. The transformed control bound as in (4.23) for (4.3) is

$$-c_d g \leq \frac{1}{M} u(t) \leq c_a g. \tag{4.28}$$

The rewritten HOCBF constraint (4.6) can only conflict with the left inequality of (4.28). Thus, following (4.24) and combining (4.6) with (4.28), the feasibility constraint is $b_{hF}(x(t)) \geq 0$, where

$$b_{hF}(x(t)) = \frac{F_r(v(t))}{M} + (p_1 + p_2)(v_p - v(t)) + p_1 p_2(z(t) - l_0) + c_d g. \tag{4.29}$$

Since $\frac{F_r(v(t))}{M} \geq 0, \forall t \geq 0$, we can replace the last equation by

$$\hat{b}_{hF}(x(t)) = (p_1 + p_2)(v_p - v(t)) + p_1 p_2(z(t) - l_0) + c_d g. \tag{4.30}$$

The satisfaction of $\hat{b}_{hF}(x(t)) \geq 0$ implies the satisfaction of $b_{hF}(x(t)) \geq 0$. Although the relative degree of $b(x(t)) = z(t) - l_0$ is two, the relative degree of $\hat{b}_{hF}(x(t))$ is only one. We then define $\hat{b}_{hF}(x(t))$ to be a CBF by choosing $\alpha_1(b(x(t))) = kb(x(t)), k > 0$ in Definition 3.3. Any control $u(t)$ should satisfy the HOCBF constraint (3.8) which in this case was given in (4.5):

$$\frac{u(t)}{M} \leq \frac{F_r(v(t))}{M} + (\frac{p_1 p_2}{p_1 + p_2} + k)(v_p - v(t)) \\ + \frac{k p_1 p_2}{p_1 + p_2}(z(t) - l_0) + \frac{k c_d g}{p_1 + p_2} \tag{4.31}$$

In order to reformulate the last equation in the form of (4.26), we try to determine an appropriate value for k above. We require $\varphi(\boldsymbol{x}(t))$ to satisfy $L_f\varphi(\boldsymbol{x}(t)) \geq 0$ as shown in one of the conditions in Theorem 4.5, thus, we wish to exclude the term $z(t) - l_0$ in $\varphi(\boldsymbol{x}(t))$ since its derivative $v_p - v(t)$ is usually negative. By equating the coefficients of the term $z(t) - l_0$ in (4.31) and (4.6), we have

$$\frac{kp_1p_2}{p_1 + p_2} = p_1p_2. \tag{4.32}$$

Thus, we get $k = p_1 + p_2$. By substituting k back into (4.31), we have

$$\frac{u(t)}{M} \leq \frac{F_r(v(t))}{M} + (p_1 + p_2)(v_p - v(t)) + p_1p_2(z(t) - l_0) + \varphi(\boldsymbol{x}(t)), \tag{4.33}$$

where

$$\varphi(\boldsymbol{x}(t)) = \frac{p_1p_2}{p_1 + p_2}(v_p - v(t)) + c_d g. \tag{4.34}$$

It is easy to check that the relative degree of the last function is one, $L_f\varphi(\boldsymbol{x}(t)) = \frac{p_1p_2}{p_1+p_2}\frac{F_r(v(t))}{M} \geq 0$ and $L_g\varphi(\boldsymbol{x}(t)) = \frac{p_1p_2}{p_1+p_2}L_gL_fb(\boldsymbol{x}(t))$. Thus, all the conditions in Theorem 4.5 are satisfied, except for $\varphi(\boldsymbol{x}(0)) \geq 0$, which depends on the initial state $\boldsymbol{x}(0)$ of system (4.2).

It follows that the single feasibility constraint $\varphi(\boldsymbol{x}(t)) \geq 0$ for the ACC problem is actually a speed constraint (following (4.34)):

$$v(t) \leq v_p + \frac{c_d g(p_1 + p_2)}{p_1 p_2}. \tag{4.35}$$

If $p_1 = p_2 = 1$ in (4.6), the condition above reduces to the requirement that the half speed difference between the front and ego vehicles should be greater than $-c_d g$ in order to guarantee the ACC problem feasibility.

It is worth pointing out that we can find other sufficient conditions such that the ACC problem is guaranteed to be feasible by choosing different HOCBF definitions (different class $\mathcal{K}$ functions) in the above process.   ∎

4.3   Feasibility Conditions for ACC

In this section, we return to Example 4.4 by considering a specific case study of the ACC problem. Another application of the feasibility constraint method to a traffic merging problem can be found in [99]. All computations and simulations involved were conducted in MATLAB. We used quadprog to solve the quadratic programs and ode45 to integrate the dynamics.

In addition to the dynamics (4.2), the safety constraint (4.4), the control bound (4.3), and the minimization of the cost $\int_0^T \left(\frac{u(t) - F_r(v(t))}{M}\right)^2 dt$, we also consider a desired speed requirement such that v approaches a desired speed $v_d > 0$. We use the relaxed CLF to implement this speed requirement, i.e., we define a CLF $V = (v - v_d)^2$, and choose $c_1 = c_2 = 1, c_3 = \epsilon > 0$ in Definition 2.4. Thus, any control input should satisfy the CLF constraint (2.16).

We consider the HOCBF constraint (4.6) to implement the safety constraint (4.4), and consider the sufficient condition (4.35) introduced in the last section to guarantee the feasibility of the ACC problem. We use a HOCBF with $m = 1$ to impose this condition, as introduced in (4.25). We define $\alpha(\cdot)$ as a linear function in (4.25).

Following the discretization method introduced at the end of Chap. 2, we partition the time interval $[0, T]$ into a set of equal time intervals $\{[0, \Delta t), [\Delta t, 2\Delta t), \ldots\}$, where $\Delta t > 0$. In each interval $[\omega \Delta t, (\omega + 1)\Delta t)$ $(\omega = 0, 1, 2, \ldots)$, we assume the control is constant (i.e., the overall control will be piecewise constant) and reformulate the ACC problem as a sequence of QPs. Specifically, at $t = \omega \Delta t$ $(\omega = 0, 1, 2, \ldots)$, we solve

$$u^*(t) = \arg \min_{u(t)} \frac{1}{2} u(t)^T H u(t) + F^T u(t) \tag{4.36}$$

$$u(t) = \begin{bmatrix} u(t) \\ \delta(t) \end{bmatrix}, \quad H = \begin{bmatrix} \frac{2}{M^2} & 0 \\ 0 & 2p_{acc} \end{bmatrix}, \quad F = \begin{bmatrix} \frac{-2F_r(v(t))}{M^2} \\ 0 \end{bmatrix}.$$

subject to

$$A_{\text{clf}} u(t) \leq b_{\text{clf}},$$

$$A_{\text{limit}} u(t) \leq b_{\text{limit}},$$

$$A_{\text{hocbf\_safety}} u(t) \leq b_{\text{hocbf\_safety}},$$

$$A_{\text{fea}} u(t) \leq b_{\text{fea}},$$

where $p_{acc} > 0$ and the constraint parameters are

$$A_{\text{clf}} = [L_g V(x(t)), \quad -1],$$
$$b_{\text{clf}} = -L_f V(x(t)) - \epsilon V(x(t)),$$

$$A_{\text{limit}} = \begin{bmatrix} 1, 0 \\ 1, 0 \end{bmatrix},$$

$$b_{\text{limit}} = \begin{bmatrix} c_a M g \\ -c_d M g \end{bmatrix},$$

$$A_{\text{hocbf\_safety}} = \left[\frac{1}{M}, 0 \right],$$

$$b_{\text{hocbf\_safety}} = \frac{F_r(v(t))}{M} + (p_1 + p_2)(v_p - v(t)) + p_1 p_2 (z(t) - l_0)$$

$$A_{\text{fea}} = \left[\frac{p_1 p_2}{M(p_1 + p_2)}, 0 \right],$$

Table 4.1 Simulation parameters for the ACC problem

Para.	Value	Units	Para.	Value	Units
$v(0)$	6	m/s	$z(0)$	100	m
v_p	13.89	m/s	v_d	24	m/s
M	1650	kg	g	9.81	m/s^2
f_0	0.1	N	f_1	5	Ns/m
f_2	0.25	Ns^2/m	l_0	10	m
Δt	0.1	s	ϵ	10	Unitless
$c_a(t)$	0.4	Unitless	$c_d(t)$	0.4	Unitless
p_{acc}	1	Unitless			

$$b_{\text{fea}} = \frac{p_1 p_2 F_r(v(t))}{M(p_1 + p_2)} + \frac{p_1 p_2}{p_1 + p_2}(v_p - v(t)) + c_d g.$$

After solving (4.36), we update (4.2) with $u^*(t)$, $\forall t \in (t_0 + \omega \Delta t, t_0 + (\omega + 1)\Delta t)$ and the simulation parameters used in this case study are listed in Table 4.1.

We first present in Fig. 4.5 a situation in which the ego vehicle exceeds the speed constraint from the feasibility constraint (4.35) resulting in the QP becoming infeasible. However, this infeasibility does not always hold since the feasibility constraint (4.35) is just a sufficient condition for the feasibility of QP (4.36). In order to show how the feasibility constraint (4.35) can be adapted to different parameters p_1, p_2 in (4.6), we vary them and compare the solutions without this feasibility sufficient condition in the simulation, as shown in Figs. 4.6 and 4.7.

It follows from Figs. 4.6 and 4.7 that the QPs (4.36) are always feasible with the feasibility constraint (4.35) under different p_1, p_2, while the QPs may become infeasible without this constraint. This validates the effectiveness of the feasibility constraint. We also notice that the ego vehicle cannot reach the desired speed v_d with the feasibility condition (4.35); this is due to the fact that we are limiting the vehicle speed with (4.35). In order to make the ego vehicle reach this desired speed, we choose p_1, p_2 such that the following constraint is satisfied.

$$v_p + c_d g \frac{(p_1 + p_2)}{p_1 p_2} \geq v_d \tag{4.37}$$

For example, the above constraint is satisfied when we select $p_1 = 0.5$, $p_2 = 1$ in this case. Then, the ego can reach the desired speed v_d, as shown by the blue curves in Fig. 4.8.

We also compare the feasibility constraint (4.35) with the minimum braking distance approach from [2]. This approach adds the minimum braking distance $\frac{0.5(v_p - v(t))^2}{c_d g}$ of the ego vehicle to the safety constraint (4.4):

$$z(t) \geq \frac{0.5(v_p - v(t))^2}{c_d g} + l_0, \quad \forall t \geq 0. \tag{4.38}$$

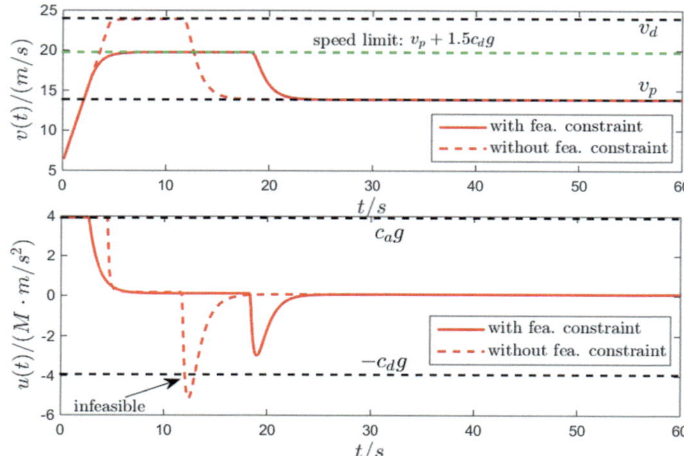

Fig. 4.5 A simple case with $p_1 = 1$, $p_2 = 2$. The QP becomes infeasible when the ego vehicle exceeds the speed limit $v_p + 1.5c_d g$ from (4.35)

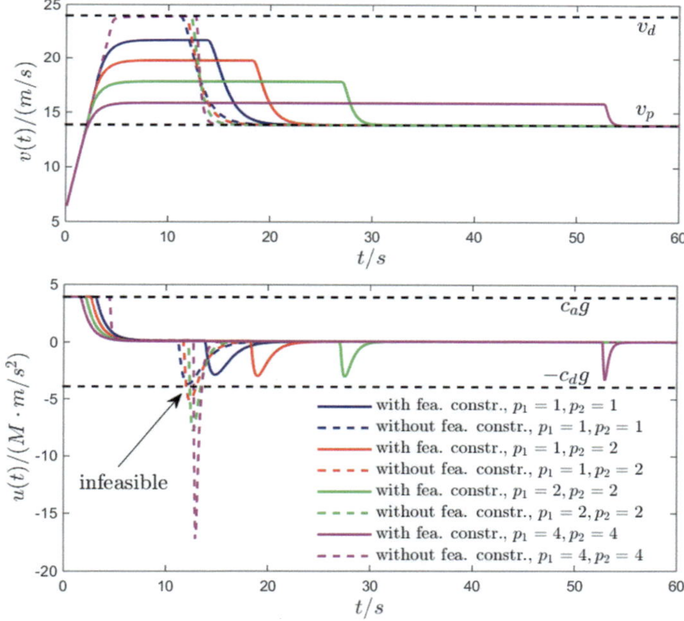

Fig. 4.6 Speed and control profiles for the ego vehicle under different p_1, p_2, with and without feasibility condition (4.35)

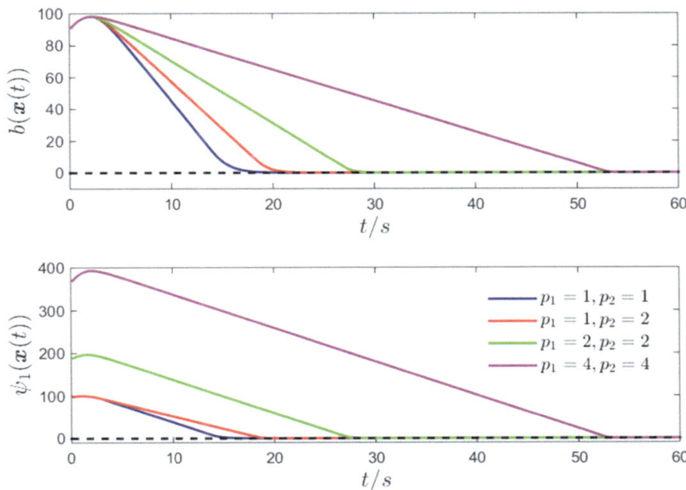

Fig. 4.7 The variation of functions $b(\boldsymbol{x}(t))$ and $\psi_1(\boldsymbol{x}(t))$ under different p_1, p_2. $b(\boldsymbol{x}(t)) \geq 0$ and $\psi_1(\boldsymbol{x}(t)) \geq 0$ implies the forward invariance of the set $C_1 \cap C_2$

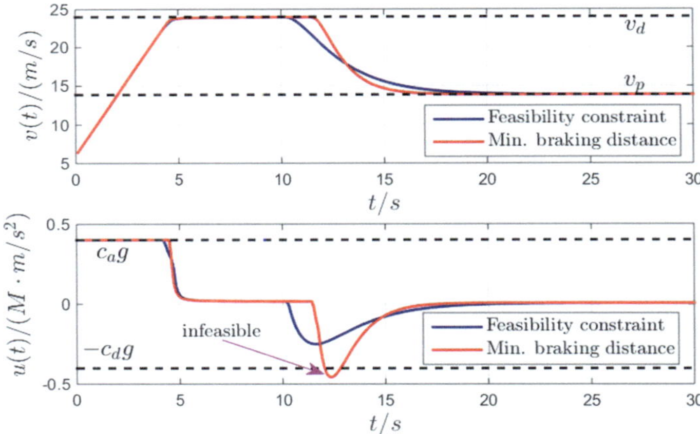

Fig. 4.8 Comparison between the feasibility constraint (4.35) with $p_1 = 0.5$, $p_2 = 1$ and the minimum braking distance approach from [2]. The HOCBF constraint for (4.38) in the minimum braking distance approach conflicts with the control bound (4.3)

Then, we can use a HOCBF with $m = 1$ (define $\alpha_1(\cdot)$ to be a linear function with slope 2 in Definition 3.3) to enforce the above constraint whose relative degree is one. As shown in Fig. 4.8, the HOCBF constraint for (4.38) conflicts with the control bounds, thus, the QP can still become infeasible. This is due to the fact that this approach adds an additional braking-distance-related constraint to the original problem, which could adversely decrease the problem feasibility as this new added constraint may conflict with existing control bounds. In contrast, the use of a proper sufficient condition for feasibility provides a novel way to make the new added feasibility constraint compliant with the existing constraints. This, therefore, can always guarantee feasibility once the sufficient conditions are determined.

Feasibility for CBF-Based Optimal Control Using Machine Learning

5

5.1 Learning CBF Parameters for Feasibility

A HOCBF constraint is *active* when a control makes it an equality at a given state. An example of how a HOCBF constraint is activated is shown in Fig. 5.1. The activation of the HOCBF will change the system trajectory in order to ensure that the system remains safe. As shown in Fig. 5.1, the precise time when a HOCBF is activated depends on how this HOCBF is designed.

In what follows, we consider multiple safety constraints instead of a single one as in (2.18). Let S denote an index set for unsafe (state) sets. System (2.11) avoids each unsafe set $j \in S$ if its state satisfies:

$$b_j(\boldsymbol{x}(t)) \geq 0, \quad \forall t \in [0, T], \quad \forall j \in S, \tag{5.1}$$

where $b_j : \mathbb{R}^n \rightarrow \mathbb{R}$ is a continuously differentiable function (not a CBF or HOCBF yet). Let S_t denote the set indexing all the unsafe set types. We define unsafe sets as being of the same "type" if they have the same geometry, meaning the conditions for problem feasibility are the same, e.g., circular unsafe sets are the same type if they have the same radius but different locations.

In the parameter learning approach we develop here, we parameterize the definition of a HOCBF (i.e., the class $\mathcal{K}$ functions in Definition 3.3) for each type of unsafe set, and aim to recursively improve a "feasibility robustness" metric introduced next.

5.1.1 Feasibility Robustness

If the HOCBF constraint corresponding to (5.1) never becomes active, then safety is guaranteed even without the use of a HOCBF. In this case, the solution space of the QP **OCP-CBF**

© The Author(s), under exclusive license to Springer Nature Switzerland AG 2023
W. Xiao et al., *Safe Autonomy with Control Barrier Functions*,
Synthesis Lectures on Computer Science,
https://doi.org/10.1007/978-3-031-27576-0_5

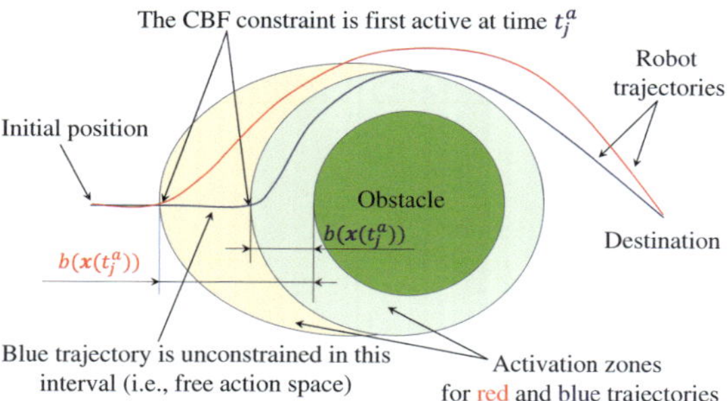

Fig. 5.1 Illustration of feasibility robustness quantification: CBF constraint activation zones depend on the dimension of the state space as state-feedback is used, but are only visualized in the $x - y$ 2D space. The CBF (HOCBF) constraint (3.8) first becomes active at time t_j^a when an agent (e.g., robot) approaches the obstacle. The CBF in the blue trajectory is active later than the one in red, thus, the robot has more freedom to move before t_j^a and it does not necessarily get close to the obstacle as shown in the figure. When $b_j(x(t_j^a))$ is minimized, the activation zone of the HOCBF is minimized, therefore, the feasibility robustness (future action space) is maximized

(2.20) is clearly at its maximum. Therefore, we only consider the case where the HOCBF constraint becomes active. Let $t_j^a \in [0, t_f]$ denote the time instant that the HOCBF constraint (3.8), corresponding to (5.1), first becomes active and (possibly) remains active afterwards if the state of system (2.11) keeps getting closer to the unsafe set. After the HOCBF constraint in the QP **OCP-CBF** (2.20) becomes active, the solution space of the QP is limited by the activated constraints, thus, the action space of the system is constrained. As an example, the HOCBF constraint along the red trajectory in Fig. 5.1 is active earlier than the blue one, therefore, the robot cannot move freely after the CBF constraint becomes active. As the solution approach of the CBF method is pointwise and myopic, our goal is to maximize the future action space of the system at each time step, i.e., to minimize the *activation zone* observed in Fig. 5.1. This is equivalent to the feasibility robustness concept we introduce here.

The feasibility robustness of a controller with respect to a constraint (5.1) can be quantified by the value of $b_j(x(t_j^a))$. The value of $b_j(x(t_j^a))$ may denote a distance metric to a (type of) unsafe set $j \in S_t$ (see the example shown in Fig. 5.1). In order to maximize the feasibility robustness (equivalently, the future action space, i.e., the future feasible control set a system can select from), we need to minimize the HOCBF value at t_j^a:

$$\min_{t_j^a} b_j(x(t_j^a)), \qquad j \in S_t. \tag{5.2}$$

The minimization of the above objective is equivalent to minimizing the activation zone of the HOCBF, hence, the feasibility robustness (or future action space) is maximized, as illustrated in Fig. 5.1.

There are three main advantages in maximizing the feasibility robustness of the controller:

1. The QPs are more likely to be feasible since fewer constraints will become active when a system gets close to a number of unsafe sets.
2. In an *unknown* environment, the controller obtained through the QPs is more robust to changes in the environment and the detection of unknown unsafe sets, since the corresponding HOCBF constraints only become active when a system gets close to these unsafe sets. If the corresponding HOCBF constraints become active before the unsafe sets are detected, the system may fail to avoid these unsafe sets.
3. There is a higher probability to find a better solution (e.g., energy optimal) if the feasibility robustness is maximized, since the QPs are less constrained. Note that proximity to an unsafe set can increase the chance of the state entering it in the presence of disturbances. However, the maximization of feasibility robustness does not mean that the system state has to get close to the unsafe sets, and this performance depends on how we design the CLF which enforces the state convergence in (2.17).

The robustness objective (5.2) depends on the time t_j^a, where t_j^a is determined once a HOCBF in the above problem is defined. Therefore, we need to consider objective (5.2) in how a HOCBF is designed. This approach requires a parameterization of a HOCBF so that we can optimize the parameter values in solving (5.2). We tackle this problem through a learning process whose objective is to maximize the feasibility robustness of the controller with respect to unknown unsafe sets as the QP-based approach for constrained optimal control problems is myopic due to the fact that the QP is solved one single time step forward. We proceed by decomposing **OCP** into two sub-problems:

- **SP1**: Optimize (2.17) which is solved with the QP-based method (2.21).
- **SP2**: Minimize the objective (5.2) after solving **SP1**.

We describe both sub-problems next, starting with **SP1**.

5.1.2   Online HOCBF and CLF-Based QP (SP1)

Sub-problem **SP1** is actually the same as the QP approach (2.21), which we re-write below for convenience replacing the CBF constraint by the more general HOCBF constraint. Thus, after partitioning the time interval $[0, T]$ into a set of equal time intervals $\{[0, \Delta t), [\Delta t, 2\Delta t), \dots\}$, where $\Delta t > 0$, we solve for the constant control $u(\omega \Delta t)$ (i.e., the overall control will be piece-wise constant) at each interval $[\omega \Delta t, (\omega + 1)\Delta t)$ ($\omega = 0, 1, 2, \dots$):

$$\min_{\boldsymbol{u}(\omega\Delta t),\delta_r(\omega\Delta t)} \mathcal{C}(||\boldsymbol{u}(\omega\Delta t)||) + p_t\delta_r^2(\omega\Delta t)$$

$$\text{s.t.} \quad L_f^m b_j(\boldsymbol{x}) + [L_g L_f^{m-1} b_j(\boldsymbol{x})]\boldsymbol{u} + O(b_j(\boldsymbol{x})) + \alpha_m(\psi_{m-1}(\boldsymbol{x})) \geq 0, \quad \forall j \in S \tag{5.3}$$

$$L_f V(\boldsymbol{x}) + L_g V(\boldsymbol{x})\boldsymbol{u} + \epsilon V(\boldsymbol{x}) \leq \delta_r$$

$$\boldsymbol{u}_{min} \leq \boldsymbol{u} \leq \boldsymbol{u}_{max}$$

As we know by now, this optimization problem can easily become infeasible. In the sequel, we show how we can use machine learning techniques to address this issue by maximizing the feasibility robustness through (5.2).

5.1.3 The Parameterization Method

The approach we follow here is similar to that of Sect. 4.1 where we modified the class $\mathcal{K}$ functions in a HOCBF through a multiplicative penalty parameter. Here, however, in order to recursively improve the feasibility of the QP (5.3), we parameterize the actual family of class $\mathcal{K}$ functions $\alpha_1(\cdot), \alpha_2(\cdot), \ldots, \alpha_m(\cdot)$ (see also [72]), where m denotes the relative degree of the constraint $b(\boldsymbol{x}) \geq 0$ in the definition of a HOCBF $b(\boldsymbol{x})$. Let $\psi_0(\boldsymbol{x}) := b(\boldsymbol{x})$. Since power functions are the most frequently used class $\mathcal{K}$ functions, we select the $\alpha_i(\cdot)$ functions in (3.3) as follows:

$$\psi_i(\boldsymbol{x}) := \dot{\psi}_{i-1}(\boldsymbol{x}) + p_i\psi_{i-1}^{q_i}(\boldsymbol{x}), \quad i \in \{1, \ldots, m\} \tag{5.4}$$

where $p_i > 0, i \in \{1, \ldots, m\}$ and $q_i \geq 1, i \in \{1, \ldots, m\}$. Then, we can obtain the HOCBF constraint (3.8) when combining it with dynamics (2.11), as shown in Definition 3.3.

Recall that S_t denotes the set that indexes all the unsafe set types. For each type of unsafe set $j \in S_t$, we consider an arbitrary location for it and obtain an unsafe set constraint $b_j(\boldsymbol{x}(t)) \geq 0$, similar to (5.1). Let $\boldsymbol{p} := (p_1, \ldots, p_m), \boldsymbol{q} := (q_1, \ldots, q_m)$. We know from [72] that the values of $\boldsymbol{p}, \boldsymbol{q}$ affect the feasible set for the decision variables of (5.3), as well as what time t_j^a the HOCBF constraint (3.8) will be active, i.e., we can rewrite $b_j(\boldsymbol{x}(t_j^a))$ as $b_j(\boldsymbol{x}(t_j^a))_{\boldsymbol{p},\boldsymbol{q}}$. Since t_j^a depends on $\boldsymbol{p}, \boldsymbol{q}$ and $b_j(\cdot)$ no longer explicitly depends on $\boldsymbol{x}(t_j^a)$ (i.e., $b_j(\boldsymbol{x}(t_j^a))_{\boldsymbol{p},\boldsymbol{q}}$ is fixed once $\boldsymbol{p}, \boldsymbol{q}$ are given), we define

$$\mathcal{D}_j(\boldsymbol{p}, \boldsymbol{q}) := b_j(\boldsymbol{x}(t_j^a))_{\boldsymbol{p},\boldsymbol{q}} \tag{5.5}$$

and reformulate (5.2) so that the minimization is over the parameter vectors $\boldsymbol{p}, \boldsymbol{q}$:

$$\min_{\boldsymbol{p},\boldsymbol{q}} \mathcal{D}_j(\boldsymbol{p}, \boldsymbol{q}), j \in S_t. \tag{5.6}$$

We can, therefore, view the minimization of $\mathcal{D}_j(\boldsymbol{p}, \boldsymbol{q})$ as the maximization of the feasibility robustness that depends on $\boldsymbol{p}, \boldsymbol{q}$. However, this optimization problem is hard to solve as t_j^a corresponds to a particular trajectory resulting from a sequence of QPs from the initial time to the final time. We will introduce a solution approach using machine learning techniques in the following section.

5.1.4 Offline Feasibility-Guided Optimization (SP2)

Note that subproblem **SP2** depends on subproblem **SP1**, while the feasibility of **SP2** depends on the control bounds (2.19). Given an arbitrary $x(0)$, one can generally expect most of the p, q values to result into infeasible solutions of problem (5.3), which makes (5.6) difficult to solve. Therefore, we need to first resolve the infeasibility problem of sub-problem **SP1**.

We proceed by randomly sampling p, q values over their domain (positive), and for each set of p, q values, we solve problem (5.3) until the terminal state constraint is satisfied within some allowed error. If problem (5.3) is feasible at all times, then we label this particular set of p, q values as $+1$, otherwise, we label it as -1. In this way, we eventually obtain sets of feasible and infeasible p, q points. Note that the penalty method described previously guarantees that $+1$ data points exist given the control bounds (2.19) if certain conditions are satisfied. We assume that the control bounds (2.19) are properly defined such that we can select balanced data sets for better classification, e.g., the same number of feasible and infeasible samplings with a large enough data size, from the randomly sampled data. Then, we can apply a standard classification method, e.g., a support vector machine (SVM) [10], to classify these two balanced sets and get a continuously differentiable hypersurface

$$\mathfrak{H}_j : \mathbb{R}^{2m} \to \mathbb{R}, \tag{5.7}$$

where

$$\mathfrak{H}_j(p, q) \geq 0 \tag{5.8}$$

denotes the set of p, q values which leads to the feasible solution of QPs (5.3), i.e., it defines the feasibility constraint for the set of p, q values associated with the QPs (5.3). With the assistance of the feasiblity classification hypersurface, we look further to optimize (5.6), i.e., we consider (5.6) subject to (5.8).

Our goal is to get the set of p, q values such that (5.8) is satisfied. However, (5.8) is usually complex as it is directly obtained from the classifier hypersurface. Therefore, we view $\mathfrak{H}_j(p, q) \geq 0$ as a constraint applied to a dynamic system with states p, q and formulate an associated HOCBF. In particular, just like $b(x)$ is associated with the dynamic system (2.11), we need to introduce an *auxiliary dynamic system* for $\mathfrak{H}_j(p, q)$ and take p, q as state variables, as shown in the sequel.

We have imposed the assumption that the control bounds (2.19) are properly defined such that the whole problem is well-posed to get a proper constraint (5.8) from the hypersurface. However, the learned hypersurface is still complex, even with the HOCBF method, and this makes the optimization problem (5.6) subject to (5.8) hard to solve. We use the following dynamic process to simplify this optimization problem.

We start at some feasible $p_0 \in \mathbb{R}^m$, $q_0 \in \mathbb{R}^m$ to search for the optimal p, q values. Since the determination of the optimal p, q is a dynamic process, we define the gradient (auxiliary dynamics) for p, q as controllable variations of p, q through the dynamic system

$$(\dot{p}(\mathsf{t}), \dot{q}(\mathsf{t})) = v(\mathsf{t}),$$
$$\text{s.t. } \mathfrak{H}_j(p, q) \geq 0, \tag{5.9}$$
$$p(\mathsf{t}_0) = p_0, q(\mathsf{t}_0) = q_0,$$

where $v \in \mathbb{R}^{2m}$ denotes an input vector in the dynamic process constructed in order to determine the optimal p, q. The variable t denotes the dynamic process time for the optimization of (5.6), which is different and independent from t in (2.11) and problem (5.3), and $\mathsf{t}_0 \in \mathbb{R}$ denotes the initial time.

Now viewing (5.8) as a state constraint analogous to $b_j(x) \geq 0$ in (5.1), we seek to treat it as a HOCBF. Considering the feasibility of problem (5.3), the dynamic process that is controlled by v should be subjected to (5.9), as well as to the HOCBF constraint for (5.8) since we define the hypersurface in (5.8) to be a HOCBF, as discussed in the last three paragraphs. Since we take all the state variables of the auxiliary dynamics (5.9) as the input for the classifier, the relative degree of the feasibility constraint (5.8) with respect to (5.9) is 1, i.e., we only need to differentiate $\mathfrak{H}_j(p, q)$ along the dynamics (5.9) once to let v show up. A control v should satisfy the HOCBF constraint (3.8) which in this case is

$$\frac{d\mathfrak{H}_j(p, q)}{d(p, q)} v + \beta_1(\mathfrak{H}_j(p, q)) \geq 0, \tag{5.10}$$

where $\beta_1(\cdot)$ is an extended class $\mathcal{K}$ function as $\mathfrak{H}_j(p, q)$ could be negative due to possible classification errors. Any control v that satisfies (5.10) implies that the resulting p, q (determined by v) leads to a feasible solution of QPs (5.3) in the dynamic process.

We enforce the feasibility constraint (5.8) by the HOCBF constraint (5.10) in which the control v explicitly shows. However, the cost function (5.6) is only defined over the state of the auxiliary dynamics (5.9), and we also wish the control v to show up in the cost function, which is required by the CBF-based optimization, as shown in Sect. 5.1.2. Therefore, we consider the derivative of the cost function (5.6) as our new cost to let v show up in the cost function. Recall that (5.6) is the robustness metric that we wish to minimize. As long as the derivative of (5.6) is negative, we can ensure that (5.6) is decreasing at each time step by discretizing t similar to sub-problem **SP1**. To modify (5.6), we proceed as follows.

By taking the derivative of (5.6) with respect to t, we have

$$\frac{d\mathcal{D}_j(p(\mathsf{t}), q(\mathsf{t}))}{d\mathsf{t}} = \frac{d\mathcal{D}_j(p(\mathsf{t}), q(\mathsf{t}))}{d(p(\mathsf{t}), q(\mathsf{t}))} v. \tag{5.11}$$

Then, we reformulate sub-problem **SP1** through the dynamic process (5.9). The result is the **Feasibility-Guided Optimization** (FGO) algorithm that is implemented by the same approach as introduced in Sect. 5.1.2, i.e., we discretize t, and at each $t = \omega\Delta t$, $\omega \in \{0, 1, \ldots\}$, where $\Delta t > 0$ denotes the discretization constant, we solve

$$\min_{\nu(t)} \frac{d\mathcal{D}_j(\boldsymbol{p}(t), \boldsymbol{q}(t))}{d(\boldsymbol{p}(t), \boldsymbol{q}(t))} \boldsymbol{\nu}(t),$$

$$\text{s.t.} \quad \frac{d\mathfrak{H}_j(\boldsymbol{p}, \boldsymbol{q})}{d(\boldsymbol{p}, \boldsymbol{q})} \boldsymbol{\nu} + \beta_1(\mathfrak{H}_j(\boldsymbol{p}, \boldsymbol{q})) \geq 0, \tag{5.12}$$

$$\boldsymbol{\nu}_{min} \leq \boldsymbol{\nu} \leq \boldsymbol{\nu}_{max}.$$

where $\boldsymbol{\nu}_{min} < \boldsymbol{0}$, $\boldsymbol{\nu}_{max} > \boldsymbol{0}$ (componentwise), $\boldsymbol{0} \in \mathbb{R}^{2m}$. Then, we update (5.9) for $t \in (\omega\Delta t, (\omega + 1)\Delta t)$ with $\boldsymbol{\nu}^*(t)$. The optimization problem (5.12) is a linear program (LP) at each time step for each initial $\boldsymbol{p}, \boldsymbol{q}$ (we need to reset t for each set of initial $\boldsymbol{p}, \boldsymbol{q}$ values). Without any constraint on $\boldsymbol{\nu}$, the LP (5.12) is ill-posed because it leads to unbounded solutions. In fact, the value of $\boldsymbol{\nu}$ determines the search step length of the FGO algorithm implemented through the LP (5.12), and we want to limit this step length. Therefore, we add limitations to $\boldsymbol{\nu}$ in the LP (5.12). Note that in the last equation, $\frac{d\mathcal{D}_j(\boldsymbol{p}(t), \boldsymbol{q}(t))}{d(\boldsymbol{p}(t), \boldsymbol{q}(t))}$ is a row vector of dimension $2m$, while $\boldsymbol{\nu}$ is a column vector of dimension $2m$. Therefore, the cost function in the last equation is a scalar function of $\boldsymbol{\nu}$.

After adding limitations to $\boldsymbol{\nu}$ in (5.12), the dynamic process's search step length will become bounded. Although there are control limitations on $\boldsymbol{\nu}$, the resulting LP from the optimization (5.12) is always feasible as the relative degree of (5.8) with respect to (5.9) is 1 [71]. We also need to evaluate $\frac{\partial \mathcal{D}_j}{\partial p_1}, \ldots, \frac{\partial \mathcal{D}_j}{\partial p_m}, \frac{\partial \mathcal{D}_j}{\partial q_1}, \ldots, \frac{\partial \mathcal{D}_j}{\partial q_m}$ at each time step, i.e., evaluate the coefficients of the cost function (5.12).

The resulting process is the FGO algorithm formulated from (5.12) to optimize $\boldsymbol{p}, \boldsymbol{q}$. For each step of the FGO algorithm, any one of the following four conditions may terminate it: (a) the problem (5.3) becomes infeasible (since the hypersurface from the classifier cannot ensure 100% accuracy), (b) the computed values of $\frac{\partial \mathcal{D}_j}{\partial p_1}, \ldots, \frac{\partial \mathcal{D}_j}{\partial p_m}, \frac{\partial \mathcal{D}_j}{\partial q_1}, \ldots, \frac{\partial \mathcal{D}_j}{\partial q_m}$ are all 0, (c) the objective function value of (5.6) is greater than the current known minimum value. (d) the iteration time exceeds some $N \in \mathbb{N}$. We present the FGO algorithm in Algorithm 1. We note that this approach is more computationally efficient than one that directly incorporates the barrier function in the cost, as Algorithm 1 only involves QPs.

If we consider (5.12) without constraint (5.10), then we have the commonly used gradient descent (GD) algorithm. The FGO algorithm is more efficient compared with GD since the solution searching path is guided by the feasibility of (5.3). However, the hypersurface in (5.8) cannot guarantee the correctness of the FGO method due to possible classification errors. We can apply GD one step forward whenever the FGO algorithm terminates to alleviate this problem.

Algorithm 1: Feasibility-Guided Optimization (FGO) algorithm

Input: Constraints (5.1), K in (2.17), system (2.11) with (2.19), N

Output: $p^*, q^*, \mathcal{D}_{min}$

1. Sample p, q in the definition of the HOCBF;
2. Discard samples that do not meet the initial conditions of HOCBF constraint (3.8);
3. Solve (5.3) for each sample for $t \in [0, T]$ and label all samples;
4. Select balanced training and testing data sets;
5. Use machine learning techniques to find classifier (5.7);
6. Pick a feasible $p_0, q_0, \mathcal{D}_{min} = \mathcal{D}_j(p_0, q_0)$, iter. $= 1$;

while *iter.$++ \leq N$* **do**

> Evaluate $\dfrac{\partial \mathcal{D}_j}{\partial p_1}, \ldots, \dfrac{\partial \mathcal{D}_j}{\partial p_m}, \dfrac{\partial \mathcal{D}_j}{\partial q_1}, \ldots, \dfrac{\partial \mathcal{D}_j}{\partial q_m}$ at p_0, q_0 with random perturbation to p_k or
> $q_k, k \in \{1, 2, \ldots, m\}$;
>
> **if** $\dfrac{\partial \mathcal{D}_j}{\partial p_k}, \dfrac{\partial \mathcal{D}_j}{\partial q_k}, \forall k \in \{1, 2, \ldots, m\}$ *is infeasible to evaluate over sub-problem (i) for all*
> *times in* $[0, T]$ **then**
>
> > Jump to the very begining of the loop;
>
> **else**
>
> > $\dfrac{\partial \mathcal{D}_j}{\partial p_k} = 0, \dfrac{\partial \mathcal{D}_j}{\partial q_k} = 0, \exists k \in \{1, 2, \ldots, m\}$ is infeasible to evaluate over sub-problem (i)
> > for all times in $[0, T]$;
>
> **end**
>
> Solve the optimization (5.12) and get new p, q;
> Solve the problem (5.3) with p, q;
> **if** *(5.3) is feasible for all $t \in [0, T]$* **then**
>
> > **if** $\mathcal{D}_{min} \geq \mathcal{D}_j(p, q)$ **then**
> >
> > > $\mathcal{D}_{min} = \mathcal{D}_j(p, q), p_0 = p, q_0 = q$;
> >
> > **else**
> >
> > > break;
> >
> > **end**
>
> **else**
>
> > Solve the optimization (5.12) without (5.10) and get new p, q;
> > Solve the problem (5.3) with p, q;
> > **if** $\mathcal{D}_{min} \geq \mathcal{D}_j(p, q)$ **then**
> >
> > > $\mathcal{D}_{min} = \mathcal{D}_j(p, q), p_0 = p, q_0 = q$;
> >
> > **else**
> >
> > > break;
> >
> > **end**
>
> **end**

end

$p^* = p_0, q^* = q_0$;

Note that we can update the training set and get a new classifier in (5.8) after running the FGO algorithm for a number of different initial samples p_0, q_0, i.e., re-initialize (5.9) for each FGO process. Once we have learned feasibility and robustness for some known types of unsafe sets with the FGO algorithm, we can use these unsafe sets to approximate other types of unsafe sets.

Remark 5.1 (*Time complexities of subproblems SP1 and SP2*). The time complexity of **SP1**, i.e., the QP (5.3), is $O(d^3)$, where $d = q + 1$ is the dimension of decision variables. Since the CBF method (after pre-training) does not need planning, it is more computationally efficient than path planning methods, such as Rapidly-exploring Randomized Trees (RRT) [29] and A* [22]. On the other hand, the time complexity of **SP2** is that of a LP [66], i.e., $O((d + c)^{1.5} d L)$, where d, c are the numbers of decision variables and constraints, respectively, and L is a given parameter. Thus, the complexity of the FGO algorithm is almost the same as the GD one, as it just has one more constraint than the GD method, hence the computational times are comparable.

5.1.5   Robot Control Using Feasibility-Guided Optimization

In this section, we apply the FGO algorithm to a robot control problem in a MATLAB setting and show several simulations results. The robot moves is an environment with several obstacles of the same type, but the number of obstacles and their locations are unknown to the robot. The robot is equipped with a sensor ($\frac{2}{3}\pi$ field of view (FOV) and $7\,\mathrm{m}$ sensing distance with $1\,\mathrm{m}$ uncertainty) to detect the obstacles.

Defining $x := (x, y, \theta, v)$, $u := (u_1, u_2)$, the dynamics are:

$$\begin{aligned}
\dot{x} &= v\cos(\theta), & \dot{y} &= v\sin(\theta), \\
\dot{\theta} &= u_1, & \dot{v} &= u_2,
\end{aligned} \tag{5.13}$$

where x, y denote the location along the x, y axis, respectively, θ denotes the heading angle of the robot, v denotes the linear speed, and u_1, u_2 denote the two control inputs for turning and acceleration, respectively.

The objective is defined to be:

$$\min_{u(t)} \int_0^{t_f} \left[u_1^2(t) + u_2^2(t) \right] dt + p_0((x(t_f) - x_d)^2 + (y(t_f) - y_d)^2). \tag{5.14}$$

In other words, we wish to minimize the energy consumption and drive the robot to a given destination $(x_d, y_d) \in \mathbb{R}^2$, i.e., drive $(x(t), y(t))$ to (x_d, y_d), $\forall t \in [t', t_f]$, for some $t' \in [0, t_f]$. The robot dynamics are not full-state linearizable [26] and the relative degree of the position (output) is 2. Therefore, we cannot directly apply a CLF. However, the robot can arrive at the destination if its heading angle θ stabilizes to the desired direction and its speed v stabilizes to a desired speed $v_0 > 0$, i.e.,

$$\theta(t) \to \arctan(\frac{y_d - y(t)}{x_d - x(t)}), \quad v(t) \to v_0, \forall t \in [0, t_f]. \tag{5.15}$$

Now, we can apply the CLF method since the relative degrees of the heading angle and speed are 1.

The unsafe sets (5.1) are defined as circular (regular) obstacles:

$$\sqrt{(x(t) - x_i)^2 + (y(t) - y_i)^2} \geq r, \quad \forall i \in S, \tag{5.16}$$

where (x_i, y_i) denotes the location of the obstacle $i \in S$, and $r > 0$ denotes the safe distance to the obstacle.

The speed and control constraints (2.19) are defined as:

$$V_{min} \leq v(t) \leq V_{max}, \tag{5.17}$$

$$u_{1,min} \leq u_1(t) \leq u_{1,max}, u_{2,min} \leq u_2(t) \leq u_{2,max}, \tag{5.18}$$

where $V_{min} = 0\,\text{m/s}$, $V_{max} = 2\,\text{m/s}$, $u_{1,max} = -u_{1,min} = 0.2\,\text{rad/s}$, $u_{2,max} = -u_{2,min} = 0.5\,\text{m/s}^2$. Other parameters are $p_0 = 1$, $\Delta t = 0.1\,\text{s}$, $\epsilon = 10$.

We set up the FGO algorithm training environment with the initial position of the robot, the location of an obstacle (with radius $6\,\text{m}$ and $r = 7\,\text{m}$) and the destination as $(5, 25\,\text{m})$, $(32, 25\,\text{m})$ and $(45\,\text{m}, (25 + \varepsilon)m)$, respectively, where $\varepsilon \in \mathbb{R}$. The initial heading angle and speed of the robot are $0°$ and V_{max}, respectively. The parameters for the FGO algorithm are $\Delta t = 0.1$, $\boldsymbol{v}_{max} = -\boldsymbol{v}_{min} = (0.1, 0.1, 0.1, 0.1)$. The map for FGO training is shown in Fig. 5.3a.

Note that the value of ε will affect the trajectory of the robot since we have a circular obstacle. If $\varepsilon = 0$, the robot will eventually stop at the equilibrium point shown in Fig. 5.3a. If $\varepsilon > 0$, the robot goes left around the obstacle as shown in Fig. 5.3a. Otherwise, the robot turns right.

We choose a very small $\varepsilon \neq 0$ in the FGO algorithm. Since the obstacle constraint (5.16) has relative degree 2 with respect to the dynamics, the HOCBF parameters are $\boldsymbol{p} = (p_1, p_2)$, $\boldsymbol{q} = (q_1, q_2)$. We collect balanced data sets (the ratio of the samplings between $+1$ and -1 labelled data is 1:1 for both training and testing sets) from the random samplings of M training and 1000 testing samples for $\boldsymbol{p}$ and $\boldsymbol{q}$ over interval $(0, 3]$ and $(0, 2]$, respectively. Note that we allow $\boldsymbol{q}$ to be sampled in $(0, 1)$ as the robot will not get too close to the circular obstacles, although the class $\mathcal{K}$ function $p_i \psi_{i-1}^{q_i}(\boldsymbol{x})$ in (5.4) is not Lipschitz continuous when $\psi_{i-1}(\boldsymbol{x}) = 0$.

The classification model is a support vector machine (SVM) with polynomial kernel of degree 7, i.e., the kernel function $k(\boldsymbol{y}, \boldsymbol{z})$ is defined as

Table 5.1 Comparisons between the GD and FGO algorithms

Items	GD	FGO							
Training sample number M		500	1000	1500	2000	2500	3000	3500	4000
Classification accuracy		0.879	0.927	0.939	0.953	0.960	0.963	0.966	0.970
Better than GD percentage		0.210	0.248	0.254	0.252	0.244	0.282	0.288	0.266
Worse than GD percentage		0.270	0.190	0.232	0.204	0.218	0.218	0.240	0.240
$\mathcal{D}_{min}/m$ (samples min.: 5.0)	4.6	4.6	4.6	4.6	4.6	4.8	4.6	4.6	4.6

$$k(\boldsymbol{y}, \boldsymbol{z}) = (c_1 + c_2 \boldsymbol{y}^T \boldsymbol{z})^7, \tag{5.19}$$

where $\boldsymbol{y}, \boldsymbol{z}$ denote input vectors for the SVM (i.e., $\boldsymbol{y} := (\boldsymbol{p}, \boldsymbol{q})$, and $\boldsymbol{z}$ becomes a constant vector of the same dimension as $\boldsymbol{y}$ after training). We set $c_1 = 0.8$, $c_2 = 0.5$, and the comparisons between FGO and GD are shown in Table 5.1 ("better/worse than GD percentage" denotes the percentage of data in the testing set that the FGO obtains a better objective value (5.2) of subproblem (ii) than the GD method).

The FGO has better performance compared with GD in finding $\mathcal{D}_{min}$ when the number of training samples M for the hypersurface (5.10) is large enough, as shown in Table 5.1. FGO and GD have almost the same computational cost, i.e., $< 0.01s$ for both. But this advantage decreases when the classification accuracy of the hypersurface (5.10) further increases, which may be due to over-fitting. One comparison example between FGO and GD search paths is shown in Fig. 5.2a, b. Note that we can combine them to get improved capability to search for $\boldsymbol{p}^*, \boldsymbol{q}^*$. If we apply the FGO method to the good results from GD, the additional improvement percentage is around 5% among all the testing samples.

We have implemented the learned optimal HOCBF parameters $(p_1^*, p_2^*, q_1^*, q_2^*) = (0.7426, 1.9745, 1.9148, 0.7024)$ in the definition of all the HOCBFs for all obstacles in a robot exploration problem in an unknown environment. We should also note that the optimal penalties and powers are not unique. All the circular obstacles are with different size to test the robustness of the penalty method with the learned optimal parameters, and are static but randomly distributed. The robot can safely avoid all the obstacles and arrive at its destination if the obstacles do not form traps such that the robot has no way to escape, as shown in Fig. 5.3b.

We also compared the CBF-based robot exploration framework with the RRT [29] and A* [22] algorithms by considering the configuration shown in Fig. 5.3b. The pre-training for

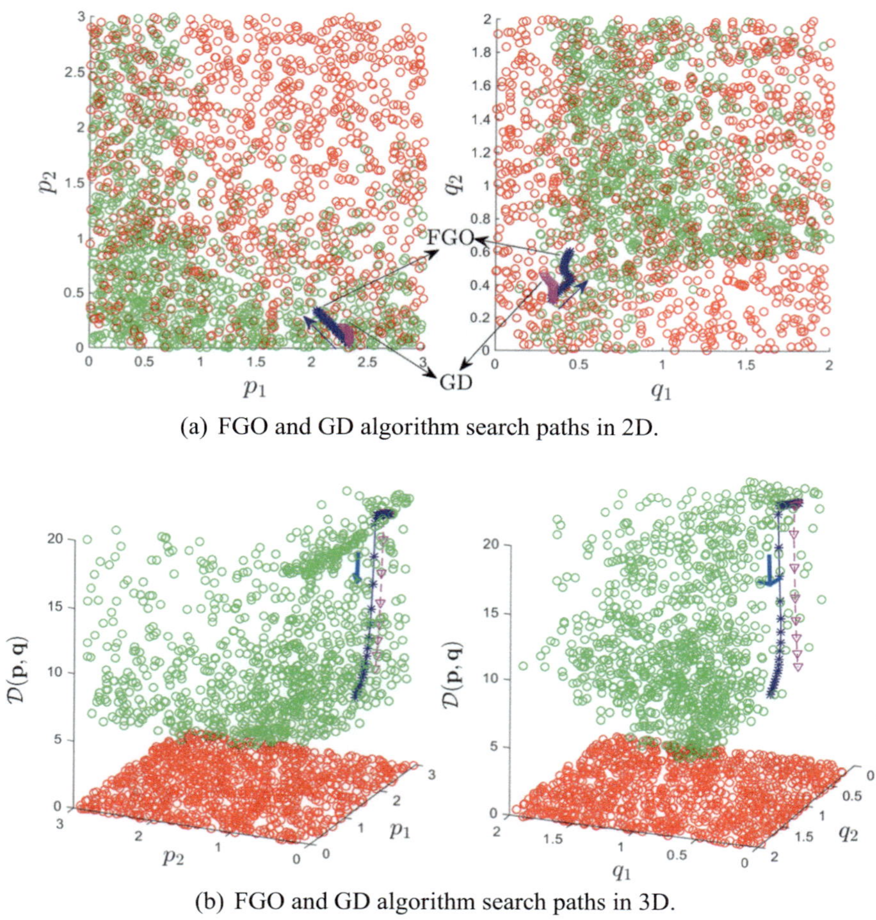

(a) FGO and GD algorithm search paths in 2D.

(b) FGO and GD algorithm search paths in 3D.

Fig. 5.2 FGO and GD comparison. The red and green circles denote infeasible and feasible points for p, q in the training samples, respectively

the CBF-based method could be several hours. Both the RRT and A* algorithms have global environment information such that they tend to choose shorter-length trajectories compared with the CBF method. But this advantage disappears if the environment is changing fast, in which case the CBF method tends to be more robust and computationally efficient. Comparisons based on four different criteria are shown in Table 5.2. In a dynamic environment, the RRT and A* algorithms need to re-plan their path at each time step, whereas the CBF method does not need to do this. Therefore, we can see that the CBF-based framework is able to better adjust to changes in the environment and is computationally efficient.

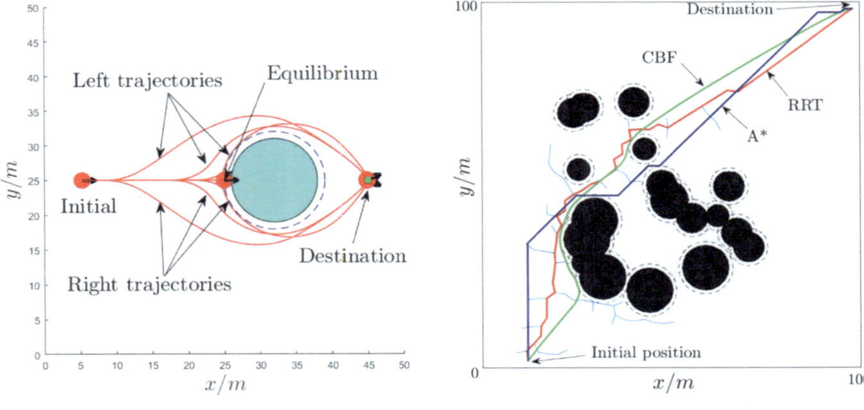

(a) FGO pre-training map with feasible example trajectories.

(b) Comparison of robot paths between CBF, A* and RRT.

Fig. 5.3 Case study setup and comparison of different planning frameworks

Table 5.2 Performance comparison between CBF, A* and RRT

Item	R.T. compute time (s)	Safety guarantee	Environment knowledge	Pre-training
CBF	<0.01	Yes	Not required	Required
A*	1.3	No	Required	Not required
RRT	0.3	No	Required	Not required

5.2   Learning Feasibility Constraints

In this section, we show how we can deal with *irregular* unsafe sets (defined below). This approach also works for regular unsafe sets, but tends to be conservative [87]. Recall that in the unknown environments we consider, the type of every unsafe set $i \in S$ that might be present is already known, and S_t denotes an index set for unsafe set types, while $S_j \subseteq S$, $j \in S_t$ denotes the index set for unsafe sets of type j.

5.2.1   Regular and Irregular Unsafe Sets

Depending on how the system initial state can affect the feasibility of the CBF-based QPs, we classify unsafe sets into two classes:

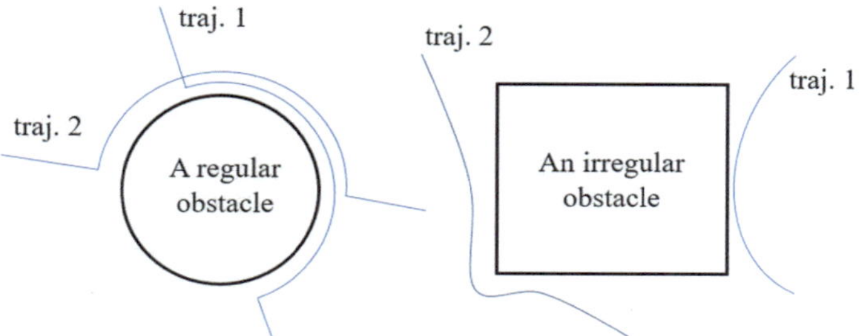

Fig. 5.4 Regular and irregular obstacle examples for an agent (e.g., robot) with nonholonomic dynamics. The agent needs the same control input effort (from the CBF-based QPs) in order to avoid the regular circular obstacle, regardless of where it is initially located, as shown in example trajectories 1 and 2. However, the agent needs a larger control input effort for trajectory 2 than the one for trajectory 1, as there are corners in the irregular rectangular obstacle. Therefore, the feasibility of the CBF-based QPs is indeed dependent on the agent initial condition (location)

Definition 5.1 (*Regular and Irregular Unsafe Sets*) Assume **OCP** is feasible. An unsafe set $C_u := \{x \in \mathbb{R} : b(x) < 0\}$ considered in the QPs (5.3) is defined as *regular* if the feasibility of all the CBF-based QPs (5.3) does not depend on the initial state $x(0)$ of system (2.11). Otherwise, we say that the set is *irregular*.

The regularity of a group of unsafe sets is checked one by one as in the QP (5.3). An irregular unsafe set generally depends on the dynamics (2.11) and corresponds to irregular shapes, such as unsafe sets with sharp corners; in such cases the system requires (locally) large control input from the CBF-based QP to avoid corners if the dynamics are nonholonomic. However, the CBF-based QPs may be feasible if the system trajectory never approaches a corner. An example of a regular unsafe set is a circular obstacle, and an example of an irregular unsafe set is a rectangle (approached by an agent with nonholonomic dynamics), as shown in Fig. 5.4.

5.2.2 Feasible and Infeasible State Sets

Recall that any QP (5.3) may be infeasible at a given state $x(t)$ at time t. Even if the system is governed by an optimal control $u^*(t)$ obtained from solving the QP (5.3), the system may still exit the constraint set formed through (5.1), since the limited control may not be able to prevent the system from leaving this set when the state approaches the set boundary, which typically happens in high relative degree systems. A simple intuitive example is a robot control problem in which the robot has limited control input (deceleration) and needs to arrive at a destination while avoiding an obstacle that is located between the robot's initial position and the destination. When the robot gets close to the obstacle with high speed, it may not be able to brake in time to avoid the obstacle since the control in QP (5.3) is limited by (2.19). However, when the speed is low, the robot can safely avoid the obstacle.

The main idea of the *sampling learning approach* in this section is to partition the state space of system (2.11) into sets in which the QP (5.3) is feasible or infeasible after a certain number of time steps. This is a difficult problem, especially for high-dimensional systems with fast dynamics. We describe next how to address this problem using machine learning techniques.

5.2.3   Sampling and Classification

Recall that the setting we consider is one where the system is in an unknown environment such that it only knows the types of the unsafe sets the environment may include, but not their number or individual locations. In order to make the learned feasibility constraint independent of the location of an unsafe set, we choose the relative coordinates $z \in \mathbb{R}^n$ between the system and unsafe set as one of the input features for machine learning techniques. For example, let the system state be $x := (x_1, x_2, \ldots, x_n)$. If x_1, x_2 denote the two-dimensional position of an object in x, then, we define input features $z := (x_1 - x_o, x_2 - y_o, x_3, \ldots, x_n)$, where $(x_o, y_o) \in \mathbb{R}^2$ denotes the two-dimensional location of the unsafe set. Along the same lines, we may also consider the relative speed and acceleration between the system and unsafe set as the input for the machine learning model in order to consider moving unsafe sets.

For each type of unsafe set $j \in S_t$, since we only consider the relative coordinates as the input for the learning model as discussed above, we arbitrarily assign a location and an orientation (if it exists) for j and randomly sample around j to find an initial state $z(0)$ in the vicinity of the unsafe set. We then solve the QP (5.3) at time 0 according to the geometry of the unsafe set:

- **Regular unsafe set**: we solve the QP (5.3) at time 0 for one time step forward.
- **Irregular unsafe set**: we solve the QP (5.3) at time 0 for $H_t \in \mathbb{N} > 1$ time steps forward.

Remark 5.2 Unlike regular unsafe sets, when dealing with irregular unsafe sets, the system may get stuck at local traps. This is why we extend the solution of the QP (5.3) to $H_t > 1$ time steps. In this case, any one of the H_t-step QPs becoming infeasible will make the system fail. The local traps can easily make the QP (5.3) infeasible, especially when the system reaches their boundary. Therefore, it is more likely to make an initial state that is located around the local traps belong to the infeasible set when we solve the QP (5.3) $H_t > 1$ time steps forward. In this way, the system may avoid the local traps if it avoids the infeasible set, and thus improve its ability to remain safe.

If the QP (or all the QPs in the case of irregular sets) (5.3) is feasible, we label the state $z(0)$ as $+1$. Otherwise, it is labelled as -1. This procedure results in two labelled classes. We employ a standard machine learning algorithm (such as Support Vector Machine (SVM), Deep Neural Network (DNN), etc.) with $z(0)$ as input to perform classification, and obtain a classification hypersurface for each $j \in S_t$ in the form:

$$H_j(z) : \mathbb{R}^n \to \mathbb{R}, \tag{5.20}$$

where $H_j(z(0)) \geq 0$ denotes that $z(0)$ belongs to the feasible set. This inequality is called the **feasibility constraint**.

Assuming the relative degree of (5.20) is m, we define the set of all controls that satisfy $H_j(z(t)) \geq 0$ as:

$$K_{fea}^{j} = \{ u \in U : L_f^m H_j(z) + L_g L_f^{m-1} H_j(z)u + O(H_j(z)) + \alpha_m(\psi_{m-1}(z)) \geq 0 \} \tag{5.21}$$

where ψ_{m-1} is recursively defined as in (3.3) by H_j with extended class $\mathcal{K}$ functions.

We define next feasibility forward invariance based on which we establish Theorem 5.1.

Definition 5.2 (*Feasibility Forward Invariance*) An optimal control problem is feasibility forward invariant for system (2.11) if its solutions starting at all feasible $x(0)$ are feasible for all $t \geq 0$.

Theorem 5.1 *Assume that the hypersurfaces $H_j(z), \forall j \in S_t$ ensure 100% feasibility and infeasibility classification accuracy. If $H_j(z(0)) \geq 0, \forall j \in S_t$, then any Lipschitz continuous controller $u(t) \in K_{fea}^{j}, \forall j \in S_t$ renders **OCP** feasibility forward invariant.*

Proof By Theorem 3.2 and $H_j(z(0)) \geq 0, \forall j \in S_t$, any control input that satisfies $u(t) \in K_{fea}^{j}, \forall j \in S_t, \forall t \in [0, \infty]$ makes $H_j(z(t)) \geq 0, \forall j \in S_t, \forall t \in [0, \infty]$. Since $H_j(z), \forall j \in S_t$ classifies the state space of system (2.11) into feasible and infeasible sub-spaces for **OCP** with 100% accuracy, it follows that this problem is feasibility forward invariant for system (2.11).

Naturally, machine learning techniques cannot ensure 100% classification accuracy. Thus, we introduce next an approach based on feedback training to improve the classification accuracy. In fact, if the classification accuracy is high enough, **OCP** may also be always feasible, since system (2.11) may never reach the infeasible part of the state space.

Similar to the sequence of QPs in (5.3), we have a feasible reformulated problem at $t = \omega \Delta t$ ($\omega = 0, 1, 2, \ldots, \frac{t_f}{\Delta t} - 1$):

$$\min_{u(\omega \Delta t), \delta_r(\omega \Delta t)} \mathcal{C}(\|u(\omega \Delta t)\|) + p_t \delta_r^2(\omega \Delta t)$$

$$\text{s.t.} \quad L_f^m b_j(x) + [L_g L_f^{m-1} b_j(x)]u + O(b_j(x)) + \alpha_m(\psi_{m-1}(x)) \geq 0, \quad \forall j \in S$$

$$L_f^m H_j(z) + [L_g L_f^{m-1} H_j(z)]u + O(H_j(z)) + \alpha_m(\psi_{m-1}(z)) \geq 0, \quad \forall j \in S \qquad (5.22)$$

$$L_f V(x) + L_g V(x)u + \epsilon V(x) \leq \delta_r$$

$$u_{min} \leq u \leq u_{max}$$

where $b(x) = b_j(x)$, and every safety constraint of the same type j uses the same $H_j(z(t)) \geq 0$, $\forall j \in S_t$.

5.2.4 Feedback Training

For each $j \in S_t$, we first sample without any hypersurface (5.20) present. After the first iteration, we obtain a hypersurface that classifies the state space of system (2.11) into feasible and infeasible sets, but with relatively low accuracy. At this point, we can add these hypersurfaces (5.20) into the QP (5.22) and sample new data points to perform a new classification leading to another classification hypersurface that replaces the old one. Iteratively, the classification accuracy is improved and the infeasible set gradually shrinks.

Since the CBF method requires the constraint to be initially satisfied, we discard the samples that do not meet this requirement. To ensure classification accuracy, we also need unbiased data samples. The workflow for this process is shown in Fig. 5.5. The infeasibility

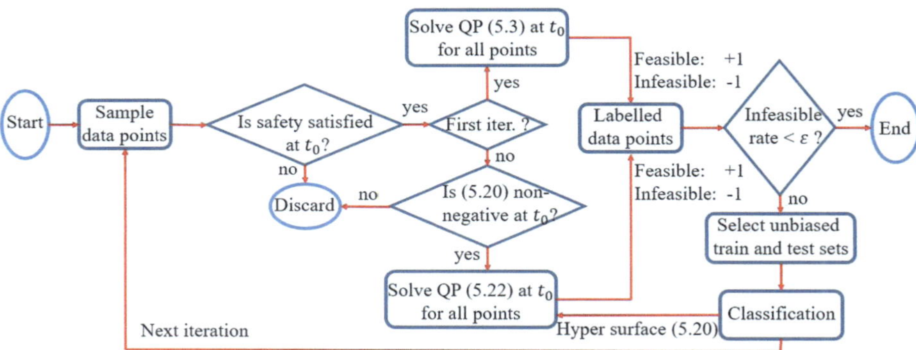

Fig. 5.5 Feedback training workflow for unsafe set $j \in S_t$ ($\varepsilon > 0$ denotes the termination threshold)

rate is the ratio of the number of infeasible samples over the total times of solving the QP (5.3) or (5.22).

5.2.5 Generalization

Since we sample data around the unsafe set, we also need to check the generalization of the hypersurface (5.20) in areas where we do not sample since system (2.11) may actually start from some state in the unsampled area. **OCP** is usually feasible when system (2.11) is far away from the unsafe set, therefore, the unsampled area should be located at the positive side of the hypersurface (5.20), which can be viewed as the generalization (i.e., not overfitting) of this hypersurface, as is usually the case in machine learning techniques.

Once we obtain a hypersurface (feasibility constraint) for a type of unsafe set in the pre-training process, we can also apply this feasibility constraint to other unsafe sets that are of the same type but with different locations, since the hypersurface only depends on the relative location of the pre-training unsafe set. This is helpful for systems in which we do not know the number and locations of the unsafe set, but we do know the type of unsafe sets that may be present in the environment.

It is also important to note that the optimal hypersurface is not unique given the training samples; this is due to the weight space symmetries [10] in neural networks.

Remark 5.3 (*Comparison between parameter and sampling learning.*) The parameter learning approach tries to learn the optimal parameters in the definition of a HOCBF such that the QP feasibility robustness, as defined in (5.2), is maximized. This can improve the adaptivity of a system in an unknown environment. However, a single set of parameters may not work for all possible initial conditions, and the system may also get stuck at local traps, so that the QP may still be infeasible. On the other hand, the sampling learning approach can deal with irregular unsafe sets as it learns a feasibility constraint by checking the feasibility of a longer than a single step receding horizon control (multi-step QP) for each sampling state. The main drawback of this approach may be the conservativeness of the learned feasibility constraint as the feasibility robustness is not considered, and this may lead to unreasonable system behaviors. This calls for a proper choice of the learning model to alleviate such conservativeness.

In addition to the learning approaches introduced in this chapter, we may also incorporate CBF-based QPs as a trainable layer (named as a BarrierNet) for neural networks [88, 91, 92], which enables end-to-end training for learning-based safety-critical control problems [90].

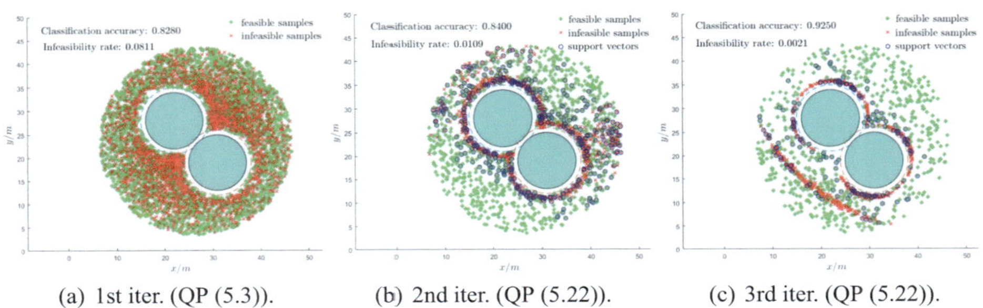

(a) 1st iter. (QP (5.3)).   (b) 2nd iter. (QP (5.22)).   (c) 3rd iter. (QP (5.22)).

Fig. 5.6 Feedback training process for the irregular obstacle. All data are sampled around the obstacle (solid circle in all sub-figures). Each sample is a four dimensional point, but is visualized in the two-dimensional $x - y$ plane

5.2.6   Robot Control by Learning Feasibility Constraints

We consider the same problem setup as in Sect. 5.1.5. We choose all class $\mathcal{K}$ functions in the definitions of all HOCBFs to be linear, and use a SVM method to classify the feasible and infeasible sets for the QP (5.3) or (5.22). We obtain a hypersurface for each type of obstacle, and apply this hypersurface to the same type obstacles with unknown locations.

We consider an irregular obstacle that is formed by two overlapped disks (with locations $(22, 28\,\text{m})$ and $(31, 19\,\text{m})$ which are unknown to the robot), as shown in Fig. 5.6. We apply the learning method introduced in Sect. 5.2.4 to recursively improve the problem feasibility, and to possibly escape from local traps. We formulate a receding horizon control of $H_t = 60$, and check the feasibility of all these H_t-step QPs. All other settings are the same as the ones in the regular obstacle case. Each training iteration is shown in Fig. 5.6.

As shown in Fig. 5.6, the classification accuracy and the infeasibility rate change similarly to the regular obstacle case in each iteration. The training results for the irregular obstacle are shown in Table 5.3. We also apply this hypersurface to the robot control problem and test for feasibility and reachability. The QPs (5.22) are always feasible on the path from the initial positions to destinations. The obstacles are safely avoided and the robot can reach its destinations. We present the results in Fig. 5.7a–b. The robot can safely avoid the local

Table 5.3 Training results for the irregular obstacle

Iter.	QP (5.22) inf. rate	Classifi. accu.	Train	Test
1	0.0811	0.8280	5000	1000 (36668)
2	0.0109	0.8400	1200	800 (89656)
3	0.0021	0.9250	1000	200 (269051)
test	0.0004			100000
gen.	0	1.0000		100000

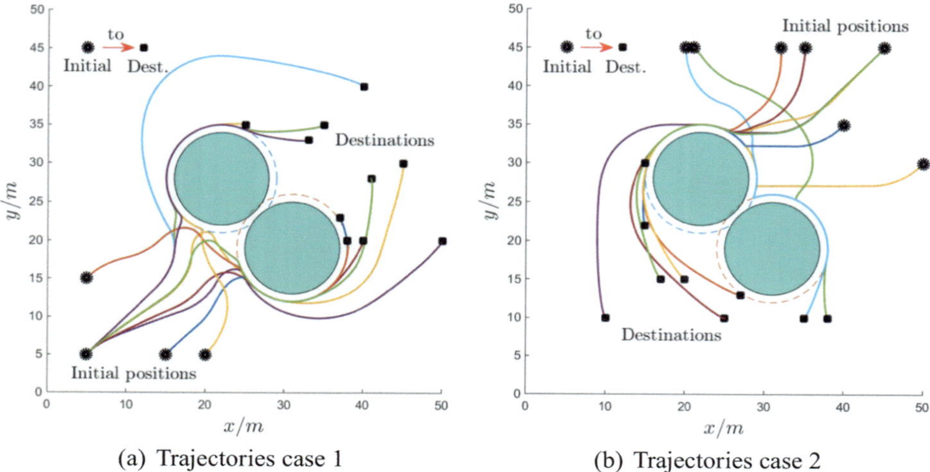

(a) Trajectories case 1   (b) Trajectories case 2

Fig. 5.7 Robot control problem feasibility and reachability test after learning in the irregular obstacle case

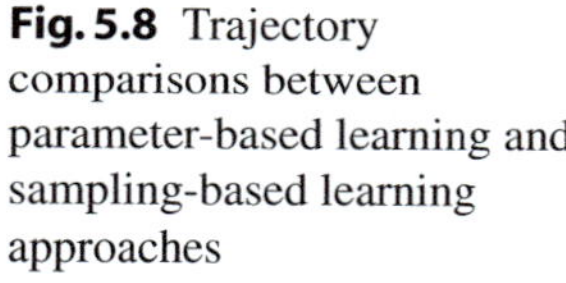

Fig. 5.8 Trajectory comparisons between parameter-based learning and sampling-based learning approaches

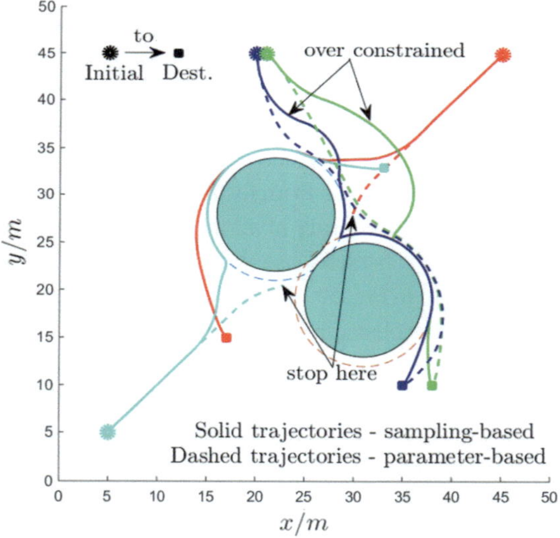

traps formed by these two circle obstacles. Thus, reachability is also improved in addition to feasibility. Finally, in Fig. 5.8 we show examples of trajectories under parameter-based learning and sampling-based learning which illustrate the differences between them. For example, we can see that the sampling-based learning method may lead to more conservative trajectories, while the parameter-based learning method may make the system get stuck at local traps.

Adaptive Control Barrier Functions 6

6.1   Adaptive Control

Traditional adaptive control is a method used by a controller that must deal with uncertain parameters in the system dynamics [25]. The foundation of adaptive control is parameter estimation, which is an integral part of system identification. Common methods of estimation include recursive least squares and gradient descent. Both of these methods provide update laws that are used to modify estimates in real-time (i.e., as the system operates). Lyapunov stability is commonly used to derive these update laws and derive convergence criteria. Projection and normalization are commonly used to improve the robustness of estimation algorithms.

Such adaptive control methods can be used in CBFs for systems with uncertain parameters as well [62]. A less conservative Robust aCBF (RaCBF) that is combined with a data-driven method has been proposed in [33] to achieve adaptive safety. Machine learning techniques have also been applied to achieve adaptive safety for systems with parameter uncertainties [19, 27].

In this chapter, we do not consider traditional adaptive control methods for CBFs to guarantee safety. This is because systems with parameter uncertainties form a subclass of systems with unknown dynamics and we separately consider the case of safety guarantees for systems with unknown dynamics in the next chapter.

6.2   Parameter-Adaptive Control Barrier Functions (PACBFs)

We begin with a simple example to motivate the need for PACBFs and to illustrate the main ideas involved in their construction.

© The Author(s), under exclusive license to Springer Nature Switzerland AG 2023
W. Xiao et al., *Safe Autonomy with Control Barrier Functions*,
Synthesis Lectures on Computer Science,
https://doi.org/10.1007/978-3-031-27576-0_6

Example 6.1 (*Simplified Adaptive Cruise Control (ACC)*) Consider a simplified version of the ACC problem in Example 4.3 with the ego (controlled) vehicle dynamics in the form:

$$
\begin{bmatrix} \dot{v}(t) \\ \dot{z}(t) \end{bmatrix} = \begin{bmatrix} 0 \\ v_p - v(t) \end{bmatrix} + \begin{bmatrix} 1 \\ 0 \end{bmatrix} u(t), \tag{6.1}
$$

where $z(t)$ denotes the distance between the preceding and the ego vehicle, $v_p > 0$, $v(t)$ denote the velocities of the preceding and ego vehicles along the lane (the velocity of the preceding vehicle is assumed constant), respectively, and $u(t)$ is the control of the ego vehicle, subject to the constraints:

$$
u_{min} \leq u(t) \leq u_{max}, \quad \forall t \geq 0, \tag{6.2}
$$

where $u_{min} < 0$ and $u_{max} > 0$ are the minimum and maximum control input, respectively. As in Example 4.3, we require that

$$
z(t) \geq l_p, \quad \forall t \geq 0. \tag{6.3}
$$

Let $x(t) := (v(t), z(t))$ and $b(x(t)) := z(t) - l_p$. The relative degree of $b(x(t))$ is $m = 2$, so we choose a HOCBF following Definition 3.3 by defining $\psi_0(x(t)) := b(x(t))$, $\alpha_1(\psi_0(x(t))) := \psi_0(x(t))$ and $\alpha_2(\psi_1(x(t))) := \psi_1(x(t))$. We then seek a control for the ego vehicle such that the constraint (6.3) is satisfied. The control $u(t)$ should satisfy (3.8) which in this case is

$$
\underbrace{0}_{L_f^2 b(x(t))} + \underbrace{-1}_{L_g L_f b(x(t))} \times u(t) + \underbrace{v_p - v(t)}_{O(b(x(t)))} + \underbrace{v_p - v(t) + z(t) - l_p}_{\alpha_2(\psi_1(x(t)))} \geq 0. \tag{6.4}
$$

Suppose we wish to minimize $\int_0^T u^2(t)dt$. We can then use the QP-based method (2.21) to solve this ACC problem. However, the HOCBF constraint (6.4) can easily conflict with $u_{min} \leq u(t)$ in (6.2) when the two vehicles get close to each other, as shown in [71]. When this happens, the QP will be infeasible. We can use the penalty method from [71] to improve the QP feasibility, i.e., we define $\psi_1(x(t)) = \dot{b}(x(t)) + pb(x(t))$, $\psi_2(x(t)) = \dot{\psi}_1(x(t)) + p\psi_1(x(t))$, $p > 0$ in (3.3). Based on (6.4), the control $u(t)$ should then satisfy:

$$
\underbrace{0}_{L_f^2 b(x(t))} + \underbrace{-1}_{L_g L_f b(x(t))} \times u(t) + \underbrace{p(v_p - v(t))}_{O(b(x(t)))} + \underbrace{p(v_p - v(t)) + p^2(z(t) - l_p)}_{p\alpha_2(\psi_1(x(t)))} \geq 0. \tag{6.5}
$$

Given u_{min}, we can find a small enough value for p such that (6.5) will not conflict with u_{min} in (6.2), i.e., the QP is always feasible. However, in practice the value of u_{min} is not a constant; it depends on weather conditions, different road surfaces, etc. Therefore, a proper choice of p for specific conditions ensuring that the QP is feasible is not easy to make when the environment changes. Moreover, the assumption of a constant speed v_p for the front vehicle is too strong, and noise in the vehicle dynamics may also make the QP infeasible.

This motivates us to define an AdaCBF that works for time-varying control bounds and noisy dynamics, i.e., it theoretically ensures that the QP is always feasible. ∎

The key idea of the PACBF method in converting a regular CBF into an adaptive one is to include the penalty terms as shown in (6.5) and then replace them by time-varying functions with suitable properties as detailed next.

Starting with a relative degree m function $b : \mathbb{R}^n \to \mathbb{R}$, let $\psi_0(x) := b(x)$. Then, instead of using a constant penalty $p_i > 0, i \in \{1, \ldots, m\}$ for each class $\mathcal{K}$ function $\alpha_i(\cdot)$ in the definition of a HOCBF [71], we define a time-varying **penalty function** $p_i(t) \geq 0, i \in \{1, \ldots, m\}$, and use it as a multiplicative factor for each class $\mathcal{K}$ function $\alpha_i(\cdot)$. Let $p(t) := (p_1(t), \ldots, p_m(t))$. Similar to (3.3), we define a sequence of functions $\psi_i : \mathbb{R}^n \times \mathbb{R}^m \to \mathbb{R}, i \in \{1, \ldots, m\}$ in the form:

$$
\begin{aligned}
\psi_1(x, p(t)) &:= \dot{\psi}_0(x) + p_1(t)\alpha_1(\psi_0(x)), \\
\psi_i(x, p(t)) &:= \dot{\psi}_{i-1}(x, p(t)) + p_i(t)\alpha_i(\psi_{i-1}(x, p(t))), \quad i \in \{2, \ldots, m\},
\end{aligned}
\tag{6.6}
$$

where $\alpha_i(\cdot), i \in \{1, \ldots, m-1\}$ is a $(m-i)$th order differentiable class $\mathcal{K}$ function, and $\alpha_m(\cdot)$ is a class $\mathcal{K}$ function. We further define a sequence of sets $C_i, i \in \{1, \ldots, m\}$ associated with (6.6) in the form:

$$
\begin{aligned}
C_1 &:= \{x \in \mathbb{R}^n : \psi_0(x) \geq 0\}, \\
C_i &:= \{(x, p(t)) \in \mathbb{R}^n \times \mathbb{R}^m : \psi_{i-1}(x, p(t)) \geq 0\}.
\end{aligned}
\tag{6.7}
$$

The remaining question is how to choose $p_i(t), i \in \{1, \ldots, m\}$ in (6.6). We require that $p_i(t) \geq 0, \forall i \in \{1, \ldots, m-1\}$, therefore we treat each $p_i(t)$ like a HOCBF, similar to the definition of $b(x) \geq 0$ in Definition 3.3. Just like $b(x)$ is associated with the dynamic system (2.11), we need to now introduce an *auxiliary dynamic system* for $p_i(t)$. Moreover, as in Definition 3.3, each penalty function $p_i(t), i \in \{1, \ldots, m-1\}$ will be differentiated $m - i$ times, while $p_m(t)$ is not differentiated.

Thus, we start by defining

$$
\pi_i(t) := (\pi_{i,1}(t), \pi_{i,2}(t), \ldots, \pi_{i,m-i}(t)) \in \mathbb{R}^{m-i}, \quad i \in \{1, \ldots, m-2\},
\tag{6.8}
$$

where $\pi_{i,j} \in \mathbb{R}, j \in \{1, \ldots, m-i\}$ are the auxiliary state variables. Next, we define $\pi_{m-1}(t) = p_{m-1}(t) \in \mathbb{R}$, which needs to be differentiated only once. Finally, we set $p_m(t) \geq 0$ as some function to be determined. Let $\pi_{i,1}(t) = p_i(t)$ in (6.6). We define input-output linearizable auxiliary dynamics for each p_i (we henceforth omit the time variable t for simplicity) through the auxiliary state π_i in the form:

$$
\begin{aligned}
\dot{\pi}_i &= F_i(\pi_i) + G_i(\pi_i)v_i, \quad i \in \{1, \ldots, m-1\}, \\
y_i &= p_i,
\end{aligned}
\tag{6.9}
$$

where y_i denotes the output, $F_i : \mathbb{R}^{m-i} \to \mathbb{R}^{m-i}$, $G_i : \mathbb{R}^{m-i} \to \mathbb{R}^{m-i}$, and $v_i \in \mathbb{R}$ denotes the control input for the auxiliary dynamics (6.9). The forms of F_i, G_i are mainly used to guarantee the non-negative property of p_i shown later. Their exact forms will determine the system performance, and the way to define them depends on the specific problem. For simplicity, we usually adopt linear forms. For example, we define $\dot{p}_{m-2} = \pi_{m-2,2}$ and $\dot{\pi}_{m-2,2} = v_{m-2}$ since we need to differentiate p_{m-2} twice as in Definition 3.3, and define $\dot{p}_{m-1} = v_{m-1}$ since we need to differentiate p_{m-1} once. We can initialize $\pi_i(0)$ to any vector as long as $p_i(0) > 0$.

An alternative way of viewing (6.9) is by defining a set of additional state variables which cause the dynamic system (2.11) to be augmented. In particular, let $\Pi := (\pi_1, \ldots, \pi_{m-1})$, $v := (v_1, \ldots, v_{m-1})$, where $v_i, i \in \{1, \ldots, m-1\}$ are the controls in the auxiliary dynamics (6.9). In order to properly define the PACBF, we augment system (2.11) with the auxiliary dynamics (6.9) as follows:

$$
\begin{bmatrix} \dot{x} \\ \dot{\Pi} \end{bmatrix} = \underbrace{\begin{bmatrix} f(x) \\ F_0(\Pi) \end{bmatrix}}_{F(x,\Pi)} + \underbrace{\begin{bmatrix} g(x) & 0 \\ 0 & G_0(\Pi) \end{bmatrix}}_{G(x,\Pi)} \begin{bmatrix} u \\ v \end{bmatrix},
\tag{6.10}
$$

where $F_0(\Pi) = (F_1(\pi_1), \ldots, F_{m-1}(\pi_{m-1}))$ and $G_0(\Pi)$ is a matrix composed by $G_i(\pi_i), i \in \{1, \ldots, m-1\}$ with dimension $\frac{m(m-1)}{2} \times (m-1)$. Moreover, $F : \mathbb{R}^{n+\frac{m(m-1)}{2}} \to \mathbb{R}^{n+\frac{m(m-1)}{2}}$, $G : \mathbb{R}^{n+\frac{m(m-1)}{2}} \to \mathbb{R}^{(n+\frac{m(m-1)}{2}) \times (q+m-1)}$.

Since p_i is a HOCBF with relative degree $m-i$ for (6.9), similar to (3.8), we define a constraint set $U_{cbf}(\Pi)$ for v:

$$
\begin{aligned}
U_{cbf}(\Pi) = \{ v \in \mathbb{R}^{m-1} : L_{F_i}^{m-i} p_i + [L_{G_i} L_{F_i}^{m-i-1} p_i] v_i + O(p_i) \\
+ \alpha_{m-i}(\psi_{i,m-i-1}(p_i)) \geq 0, \forall i \in \{1, 2, \ldots, m-1\} \},
\end{aligned}
\tag{6.11}
$$

where $\psi_{i,m-i-1}(\cdot)$ is defined similar to (3.3).

Definition 6.1 (*Parameter Adaptive Control Barrier Function (PACBF)*) Let $C_i, i \in \{1, \ldots, m\}$ be defined by (6.7), $\psi_i(x, p), i \in \{1, \ldots, m\}$ be defined by (6.6), and the auxiliary dynamics be defined by (6.9). A function $b : \mathbb{R}^n \to \mathbb{R}$ is a *Parameter Adaptive Control Barrier Function* (PACBF) with relative degree m for (2.11) if every $p_i, i \in \{1, \ldots, m-1\}$, is a HOCBF with relative degree $m-i$ for the auxiliary dynamics (6.9), and there exist $(m-i)$th order differentiable class $\mathcal{K}$ functions $\alpha_i, i \in \{1, \ldots, m-1\}$, and a class $\mathcal{K}$ function α_m such that

$$\sup_{u\in U, v\in U_{cbf}} [L_f^m b(x) + [L_g L_f^{m-1} b(x)]u + \sum_{i=1}^{m-1} \alpha_i(\psi_{i-1})[L_{G_i} L_{F_i}^{m-i-1} p_i]v_i$$

$$+ O(b(x), p) + \sum_{i=1}^{m-1} [L_{F_i}^{m-i} p_i]\alpha_i(\psi_{i-1}) + p_m \alpha_m(\psi_{m-1})] \geq 0, \tag{6.12}$$

for all $x \in C_1, (x, p) \in C_2 \cap, \ldots, \cap C_m$, and all $p_m \geq 0$. In (6.12), $O(b(x), p)$ denotes the remaining Lie derivative terms of $b(x)$ (or p) along f (or F_i, $i \in \{1, \ldots, m-1\}$) with degree less than m (or $m - i$), similar to the form in (3.8).

In (6.12), the explicit terms $O(b(x), p)$ are omitted for simplicity. The complex PACBF constraint (6.12) can be simplified (similar to (3.8)) if we just consider the augmented dynamics (6.10) in the form:

$$\sup_{u\in U, v\in U_{cbf}} [L_F^m b(x) + [L_G L_F^{m-1} b(x)]u + O(b(x), p, v) + \alpha_m(\psi_{m-1}(x, p))] \geq 0, \tag{6.13}$$

where $O(b(x), p, v)$ is linear in v and contains $\alpha_i(\cdot), i \in \{1, \ldots, m-1\}$ (similar to the form in (3.8)). Note that the $\alpha_i(\cdot)$ in the last equation denote general class $\mathcal{K}$ functions, and they can be different from the ones in (6.12).

Given a PACBF $b(x)$, we consider all control values $(u, v) \in U \times U_{cbf}(\Pi)$ that satisfy:

$$K_{acbf}(x, \Pi) = \{(u, v) \in U \times U_{cbf}(\Pi) : [L_g L_f^{m-1} b(x)]u$$

$$+ \sum_{i=1}^{m-1} \alpha_i(\psi_{i-1})[L_{G_i} L_{F_i}^{m-i-1} p_i]v_i + \sum_{i=1}^{m-1} [L_{F_i}^{m-i} p_i]\alpha_i(\psi_{i-1}) \tag{6.14}$$

$$+ L_f^m b(x) + O(b(x), p) + p_m \alpha_m(\psi_{m-1}) \geq 0\}.$$

Theorem 6.1 ([77]) *Given a PACBF $b(x)$ from Definition 6.1 with the associated sets $C_1, C_2, \ldots, C_m$ defined by (6.7), if $x(0) \in C_1$ and $(x(0), p(0)) \in C_2 \cap, \ldots, \cap C_m$, then any Lipschitz continuous controller $(u(t), v(t)) \in K_{acbf}(x(t), \Pi(t))$, $\forall t \geq 0$ renders the set C_1 forward invariant for system (2.11) and $C_2 \cap \cdots \cap C_m$ forward invariant for systems (2.11), (6.9), respectively.*

Proof If $b(x)$ is a PACBF, then $p_m(t) \geq 0, \forall t \geq 0$. Constraint (6.12) is the Lie derivative form of $\dot{\psi}_{m-1}(x, p) + p_m \alpha_m(\psi_{m-1}(x, p)) \geq 0$. If $p_m(t) > 0, \forall t \geq 0$, it follows from Theorem 3.2 that $\psi_{m-1}(x(t), p(t)) \geq 0, \forall t \geq 0$. If $p_m(t) = 0$, then $\dot{\psi}_{m-1}(x, p) \geq 0$. Since $(x, p) \in C_m$ (i.e., $\psi_{m-1}(x, p) \geq 0$ is initially satisfied), we have $\psi_{m-1}(x(t), p(t)) \geq$

$0, \forall t \geq 0$. Because $p_i, \forall i \in \{1, \ldots, m-1\}$ is a HOCBF for the auxiliary dynamics (6.9), it follows from Theorem 3.2 that $p_i(t) \geq 0, \forall t \geq 0, \forall i \in \{1, \ldots, m-1\}$. Then, we can recursively prove that $\psi_i(\boldsymbol{x}(t), \boldsymbol{p}(t)) \geq 0, \forall t \geq 0, \forall i \in \{2, \ldots, m-2\}$ similarly to the case $i = m-1$, and eventually prove that $\psi_0(\boldsymbol{x}(t)) \geq 0, \forall t \geq 0$, i.e., $b(\boldsymbol{x}(t)) \geq 0, \forall t \geq 0$. Therefore, the sets C_1 and $C_2 \cap, \ldots, \cap C_m$ are forward invariant. $\qquad\square$

Remark 6.1 (*Adaptivity of PACBFs*) In the PACBF constraint (6.12), the control $\boldsymbol{u}$ of system (2.11) depends on the controls $v_i, \forall i \in \{1, \ldots, m-1\}$ of the auxiliary dynamics (6.9). The control v_i is only constrained by the HOCBF constraint in (6.11) since we require that p_i is a HOCBF, and there are no control bounds on v_i. Therefore, we partially relax the constraints on the control input of system (2.11) in the PACBF constraint (6.12) by allowing the penalty function $p_i(t), \forall i \in \{1, \ldots, m\}$ to change through $\boldsymbol{v}$. However, the forward invariance of the set C_1 is still guaranteed, i.e., the original constraint $b(\boldsymbol{x}) \geq 0$ is guaranteed to be satisfied. This is how a PACBF provides "adaptivity". Note that we may not need to define a penalty function p_i for every class $\mathcal{K}$ function $\alpha_i(\cdot)$ in (6.6); we can instead define penalty functions for only some of them.

Remark 6.2 As a practical matter, it is worth noting that the functions $p_i(t), i \in \{1, \ldots, m\}$ should be upper-bounded in order to guarantee the system safety during the discretized time implementation of the method. This is to ensure that $p_i(t)\alpha_i(\psi_{i-1}(\boldsymbol{x}, \boldsymbol{p}))$ approaches 0 as $\psi_{i-1}(\boldsymbol{x}, \boldsymbol{p})$ approaches 0. For a problem that is intrinsically feasible, we may find such upper bounds, although this is a topic requiring further study. Moreover, if these upper bounds are large, the inter-sampling effect in the solution of the QPs (2.21) becomes crucial, since even a small change of the state $\boldsymbol{x}$ could result in a large variation of $p_i(t)\alpha_i(\psi_{i-1}(\boldsymbol{x}, \boldsymbol{p}))$. As a result, we have to choose a small enough discretization time Δt in order to address the inter-sampling effect.

6.2.1   Adaptivity to Changing Control Bounds and Noisy Dynamics

As we have seen, in the HOCBF method the QPs (2.21) may be infeasible in the presence of both control limitations (2.19) and the HOCBF constraint (3.8). There are two reasons for a QP problem to become infeasible: (*i*) the control limitations (2.19) are too tight or they are time-varying such that the HOCBF constraint (3.8) will conflict with (2.19) after it becomes active; (*ii*) the dynamics (2.11) are not accurately modeled, i.e., there may be uncertain variables, etc. In case (*ii*), the HOCBF constraint (3.8) might also conflict with (2.19) when both of them become active. This is because the state variables also show up in the HOCBF constraint (3.8), thus, the noisy dynamics can easily (and randomly) change the HOCBF constraint (3.8) through the (noisy) state variables such that (3.8) may conflict with the control limitations when they are also active. However, the QP feasibility is not only improved, but in fact guaranteed as will be shown in Theorem 6.3. This is because in the PACBF method the control $\boldsymbol{u}$ in the PACBF constraint (6.12) is relaxed by $v_i, \forall i \in \{1, \ldots, m-1\}$, as discussed in Remark 6.1.

> **Theorem 6.2** ([77]) *Given a PACBF $b(\boldsymbol{x})$ from Definition 6.1 with the associated sets $C_1, \ldots, C_m$ defined by (6.7), if $b(\boldsymbol{x}(0)) > 0$, then the satisfaction of the PACBF constraint (6.12) is a necessary and sufficient condition for the satisfaction of the original constraint $b(\boldsymbol{x}) > 0$.*

Proof If $b(\boldsymbol{x}(0)) > 0$, it follows from Theorem 6.1 that we can always choose proper class $\mathcal{K}$ functions (linear, quadratic, etc.) such that $\psi_0(\boldsymbol{x}(t)) = b(\boldsymbol{x}) > 0$ and $\psi_i(\boldsymbol{x}(t), \boldsymbol{p}(t)) > 0, i \in \{1, \ldots, m-1\}, \forall t \geq 0$. Thus, the satisfaction of the PACBF constraint (6.12) is a sufficient condition for the satisfaction of the original constraint $b(\boldsymbol{x}) > 0$.

Since $b(\boldsymbol{x}) > 0$ and $\alpha_1(b(\boldsymbol{x})) > 0$, if $b(\boldsymbol{x})$ reaches zero before (faster than) $\dot{b}(\boldsymbol{x})$ (< 0) does, then the system is not safe by Nagumo's theorem [42] (i.e., $b(\boldsymbol{x}(t)) > 0, \forall t \geq 0$ is not satisfied). Therefore, $b(\boldsymbol{x})$ should reach zero after (slower than) $\dot{b}(\boldsymbol{x})$ reaches zero, and there exists an upper bound for $\frac{\dot{b}(\boldsymbol{x})}{-\alpha_1(b(\boldsymbol{x}))}$ for some $\alpha_1(\cdot)$ (as X and U are closed sets). Thus, if $b(\boldsymbol{x}(t)) > 0$, there exists a penalty function $p_1(t) \geq 0$ (since $p_1(t)$ is a HOCBF) such that $\dot{b}(\boldsymbol{x}) > -p_1(t)\alpha_1(b(\boldsymbol{x}))$ for any $\dot{b}(\boldsymbol{x})$ with respect to dynamics (2.11). With $\psi_0(\boldsymbol{x}) = b(\boldsymbol{x})$ in (6.6), we have $\dot{\psi}_0(\boldsymbol{x}) + p_1(t)\alpha_1(\psi_0(\boldsymbol{x})) > 0$ (i.e., $\psi_1(\boldsymbol{x}, \boldsymbol{p}) > 0$). The ith derivative of $b(\boldsymbol{x})$ shows up in $\psi_i, i \in \{2, \ldots, m-1\}$, and we can also prove similarly that there exists a penalty function $p_i(t) \geq 0$ (since $p_i(t)$ is a HOCBF) such that $\psi_i(\boldsymbol{x}, \boldsymbol{p}) > 0, i \in \{2, \ldots, m-1\}$ in a recursive way. Eventually, there exists $p_m(t) \geq 0$ such that $\dot{\psi}_{m-1}(\boldsymbol{x}, \boldsymbol{p}) + p_m(t)\alpha_m(\psi_{m-1}(\boldsymbol{x}, \boldsymbol{p})) \geq 0$ (i.e., $\psi_m(\boldsymbol{x}, \boldsymbol{p}) \geq 0$). Since $\psi_m(\boldsymbol{x}, \boldsymbol{p}) \geq 0$ is equivalent to the satisfaction of the PACBF constraint (6.12), it follows that the satisfaction of the PACBF constraint (6.12) is a necessary condition for the satisfaction of the original constraint $b(\boldsymbol{x}(t)) > 0$. $\qquad\square$

Note that we exclude $b(\boldsymbol{x}) = 0$ in Theorem 6.2. The satisfaction of the PACBF constraint (6.12) is equivalent to the satisfaction of $b(\boldsymbol{x}) \geq 0$, hence the system performance is not reduced by the mapping of a constraint from state to control (this is how the standard CBFs work). In essence, we have added some extra degrees of freedom to our ability to satisfy the constraints of the original optimal control problem by augmenting system (2.11)–(6.10). These degrees of freedom come from Π which is controlled by $\boldsymbol{v}$. Therefore, if the PACBF is a valid CBF for the augmented system, it guarantees the satisfaction of the original constraint and the QP feasibility. The use of a PACBF provides flexibility in constraint satisfaction in a dynamic (time-varying) way at the expense of dealing with additional state variables Π.

Example 6.2 (*Example* 6.1 *revisited*) For the SACC problem introduced in Example 6.1, we define a PACBF with $m = 2$ for (6.3). We still choose $\alpha_1(b(x(t))) = b(x(t))$ and $\alpha_2(\psi_1(x(t))) = \psi_1(x(t))$ in Definition 6.1. Suppose we only consider a penalty function $p_1(t)$ on the class $\mathcal{K}$ function $\alpha_1(\cdot)$ and define linear dynamics for p_1 in the simple form $\dot{p}_1 = v_1$ (note that $\pi_1 = p_1$ in (6.9)). In order for $b(x(t)) := z(t) - l_p$ to be a PACBF for (6.1), a control input $u(t)$ should satisfy

$$
\underbrace{0}_{L_f^2 b(x(t))} + \underbrace{-1}_{L_g L_f b(x(t))} \times u(t) + \underbrace{(z(t) - l_p)}_{[L_{G_1} p_1(t)]\alpha_1(b(x(t)))} \times v_1(t)
$$

$$
+ \underbrace{p_1(t)(v_p - v(t))}_{O(b(x(t)), p(t))} + \underbrace{v_p - v(t) + p_1(t)(z(t) - l_p)}_{\alpha_2(\psi_1(x(t), p(t)))} \geq 0.
$$
(6.15)

Comparing the above equation with (6.5), we replace the constant p in (6.5) with $p_1(t)$, therefore, the control $u(t)$ is relaxed by $v_1(t)$. This can actually guarantee the problem feasibility as will be shown in Theorem 6.3.

Since $u(t)$ depends on v_1 which has no bounds, the control input $u(t)$ in the above PACBF constraint is relaxed. Thus, this constraint is adaptive to the change of the control bound u_{min} in (6.2) and uncertainties in v_p and $x_p(t)$ from the front vehicle. Note that $p_1(t)$ should be a HOCBF for the auxiliary dynamics $\dot{p}_1 = v_1$. The control input v_1 is subject to the corresponding HOCBF constraint such that $p_1(t) \geq 0, \forall t \geq 0$ is satisfied.　■

6.2.2   Optimal Control with PACBFs

Let us return to the optimal control problem (2.17) (omitting the terminal cost) for system (2.11):

$$
\min_{u(t)} \int_0^T \mathcal{C}(\|u(t)\|)dt,
$$
(6.16)

where $\| \cdot \|$ denotes the 2-norm of a vector, $\mathcal{C}(\cdot)$ is a strictly increasing function of its argument, and $T > 0$. Suppose system (2.11) is not accurately modeled and may also have noisy states (e.g., as an additive term to (2.11)). In addition, system (2.11) has time-varying control bounds defined as $u_{min}(t), u_{max}(t) \in \mathbb{R}^q$, where we assume that $u_{min}(t), u_{max}(t)$ are Lipschitz continuous:

$$
U(t) := \{u \in \mathbb{R}^q : u_{min}(t) \leq u \leq u_{max}(t)\}.
$$
(6.17)

Assume a (safety) constraint $b(x) \geq 0$ with relative degree m has to be satisfied by system (2.11). We use the PACBF method to guarantee $b(x) \geq 0$ so that u should satisfy the PACBF constraint (6.12). Moreover, each $v_i, i \in \{1, \ldots, m - 1\}$ is constrained by the HOCBF constraint (6.11) corresponding to the constraint $p_i(t) \geq 0$ for the auxiliary dynamics (6.9).

Note that the control v from the auxiliary dynamics is only subject to the HOCBF constraints defined in (6.11). In particular, for each v_i, $i \in \{1, \dots, m-1\}$ this constraint is one-sided, i.e.,

$$v_i \geq \frac{-L_{F_i}^{m-i} p_i - O(p_i) - \alpha_{m-i}(\psi_{i,m-i-1}(p_i))}{L_{G_i} L_{F_i}^{m-i-1} p_i}, \tag{6.18}$$

if $L_{G_i} L_{F_i}^{m-i-1} p_i > 0$. The adaptivity of a PACBF depends on the auxiliary dynamics (6.9) through v_i. If v_i changes too fast, it can affect the smoothness of the control u obtained through solving the associated QPs, which may adversely affect the performance of system (2.11).

If we add control bounds on v_i, the problem feasibility may be decreased (i.e., the adaptivity of a PACBF is weakened). If we add a quadratic penalty term v_i^2 into the cost function, then p_i may maintain a large value, which contradicts the penalty method from [71] (i.e., we wish to have a small enough value of p_i to improve the problem feasibility). Therefore, in order to decrease p_i when it is large, we seek to minimize v_i and stabilize each $p_i(t)$ to a small enough value $p_i^* > 0$ (for example, as recommended by the penalty method from [71] or by the optimal penalties learned in [74]). We choose smaller p_i^* if $\alpha_i(\cdot)$ is a high-order (e.g., polynomial) function as its value is larger and requires more penalization. The choice of p_i^* can provide conditions such that the problem feasibility is guaranteed, as shown in Theorem 6.3.

Assuming the auxiliary dynamics (6.9) are input-output linearized (otherwise, we perform input-output linearization), we can use either the tracking control from [26] or a CLF to stabilize $p_i(t)$, i.e., if $m = 1$, we minimize $(p_1 - p_1^*)^2$; if $m = 2$, we define a CLF $V_1(p_1) := (p_1 - p_1^*)^2$; and if $m > 2$, we find a desired state feedback form $\hat{p}_{i,m-i}$ for $p_{i,m-i}$:

$$\hat{p}_{i,m-i} = \begin{cases} -k_1(p_i - p_i^*), & i = m-2, \\ -k_1(p_i - p_i^*) - k_2 p_{i,2} - \dots, -k_{m-i-1} p_{i,m-i-1}, \\ & i < m-2, \end{cases} \tag{6.19}$$

where $k_1 > 0, \dots, k_{m-i-1} > 0$. In the last equation, if $i = m-1$, we can directly define a CLF $V_i(\pi_i) := (p_i - p_i^*)^2$.

Then, we can define a CLF $V_i(\pi_i) := (p_{i,m-i} - \hat{p}_{i,m-i})^2$, $i \in \{1, \dots, m-1\}$ (the relative degree of $V_i(\pi_i)$ is one) to stabilize each p_i so that any control input v_i should satisfy:

$$L_{F_i} V_i(\pi_i) + L_{G_i} V_i(\pi_i) v_i + \epsilon V_i(\pi_i) \leq \delta_i \tag{6.20}$$

where δ_i is a relaxation variable that we seek to minimize.

In all cases, we may also want to stabilize p_m which is not differentiated, by minimizing $(p_m - p_m^*)^2$. Therefore, letting $\delta := (\delta_1, \delta_2, \dots, \delta_{m-1})$, we can reformulate the cost (6.16) to incorporate the PACBF as follows:

$$\min_{\boldsymbol{u}(t),\boldsymbol{v}(t),\boldsymbol{\delta}(t),p_m(t)} \int_0^T \left[\mathcal{C}(\|\boldsymbol{u}(t)\|) + \sum_{i=1}^{m-1} W_i v_i(t) \right.$$
$$\left. + \sum_{i=1}^{m-1} P_i \delta_i^2(t) + Q(p_m(t) - p_m^*)^2 \right] dt \tag{6.21}$$

subject to (2.11), (6.17), (6.12), (6.9), the HOCBF constraint in (6.11) for each $p_i \geq 0$, $i \in \{1, \dots, m-1\}$, $p_m(t) \geq 0$, and the CLF constraint (6.20). In (6.21), $W_i > 0$, $P_i > 0$, $i \in \{1, \dots, m-1\}$, and $Q \geq 0$. We can then use the QP-based approach (2.21) from Sect. 2.5 to solve (6.21).

Finding the weights in (6.21) may not be a simple task, as the ranges of the associated controllable variables may vary widely. Therefore, we normalize each term and form a convex combination:

$$\min_{\boldsymbol{u}(t),\boldsymbol{v}(t),\boldsymbol{\delta}(t),p_m(t)} \int_0^T \left[c_0 \frac{\mathcal{C}(\|\boldsymbol{u}(t)\|)}{u_{\lim}} + \sum_{i=1}^{m-1} W_i \frac{v_i(t)}{v_{i,\max}} \right.$$
$$\left. + \sum_{i=1}^{m-1} P_i \frac{\delta_i^2(t)}{\delta_{i,\max}^2} + Q \frac{(p_m(t) - p_m^*)^2}{p_{m,\max}^2} \right] dt \tag{6.22}$$

subject to the same constraints as (6.21), where $c_0 + \sum_{i=1}^{m-1} W_i + \sum_{i=1}^{m-1} P_i + Q = 1$. The upper bounds above are chosen so that $u_{\lim} = \sup_{\boldsymbol{u} \in U} \mathcal{C}(\|\boldsymbol{u}(t)\|)$, $v_{i,\max} > 0$, $\delta_{i,\max} > 0$, $p_{m,\max} > 0$. Since $v_i(t)$, $\delta_i(t)$, $p_m(t)$ do not have any natural upper bounds, it is hard to determine $v_{i,\max}$, $\delta_{i,\max}$, $p_{m,\max}$. However, we can exploit the fact that $v_{i,\max}$, $\delta_{i,\max}$, $p_{m,\max}$ are always desired to have small values and set these to be the largest acceptable values, depending on the problem of interest.

The time complexity of each QP used to solve this problem is polynomial in the dimension $d > 0$ of the decision variables. In the HOCBF-based QP, $d = q$, where q is the dimension of the control $\boldsymbol{u}$. However, in (6.22), $d = q + 2m - 1$. Thus, we see that we increase the adaptivity of the CBF method at the expense of more computation time; however, the solution of the PACBF-based QP is still fast enough.

Due to the point-wise nature of the QP-based solution method (2.21) for (6.22), we can state that an optimal control problem is "intrinsically" infeasible starting at some time instant when the safety constraint $b(\boldsymbol{x}) \geq 0$ conflicts with the control bounds (6.17) no matter how we choose the control law for system (2.11). The PACBF constraint (6.12) is active when $(\boldsymbol{u}, \boldsymbol{v})$ makes both (6.12) and (6.11) equalities. If the PACBF constraint (6.12) becomes active earlier, a QP used in the solution of (6.22) is easier to be feasible as system (2.11) has a longer time horizon to adjust its state under control bounds (6.17). Let $t_{fea} \geq 0$ denote the last time that the problem (6.16), subject to (6.17) and $b(\boldsymbol{x}) \geq 0$, is feasible. In the following theorem, we provide conditions such that the feasibility of the QPs associated with (6.22) is guaranteed.

> **Theorem 6.3** ([77]) *Suppose $x(0)$ is not on the boundary of C_1. If the PACBF constraint (6.12) is active before t_{fea}, then the feasibility of each QP associated with (6.22) is guaranteed.*

Proof Since $x(0)$ is not on the boundary of C_1, we have by Theorem 6.2 that the satisfaction of the PACBF constraint (6.12) is a necessary and sufficient condition for the satisfaction of $b(x) \geq 0$, thus, $b(x) \geq 0 \Leftrightarrow$ (6.12), and the mapping of a constraint from the state onto the control will not limit the control that system (2.11) can take with respect to the original problem (6.16), subject to (6.17) and $b(x) \geq 0$. As the problem (6.16), subject to (6.17) and $b(x) \geq 0$, is feasible, there exists a control law such that the problem is feasible after the PACBF constraint (6.12) becomes active, hence the feasibility of any QP associated with (6.22) is guaranteed. $\qquad\square$

Remark 6.3 In order to apply Theorem 6.3, we may try to find $\bar{p}_i > 0, i \in \{1, \ldots, m\}$ with $p_i^* = \bar{p}_i$ in (6.20) such that system (2.11) should take $u(t) = u_{\max}(t)$ or $u_{\min}(t)$ (or a mixture of maximum and minimum controls), $\forall t \geq t_a$ to make the QP (6.22) feasible, where $t_a \geq 0$ denotes the time that the PACBF constraint (6.12) first becomes active. Then, any $p_i^* \leq \bar{p}_i, \forall i \in \{1, \ldots, m\}$ can make a QP for (6.22) feasible with the PACBF method, as smaller penalties make the PACBF constraint become active earlier [71]. When noise is involved and it is bounded, we can apply the worst-case noise (i.e., its bound) to system (2.11) and, as above, find the $\bar{p}_i > 0$ with $p_i^* = \bar{p}_i$ in (6.20) such that system (2.11) should take $u(t) = u_{\max}(t)$ or $u_{\min}(t)$ (or a mixture of maximum and minimum controls), $\forall t \geq t_a$ to make any QP in (6.22) feasible. Then any $p_i^* \leq \bar{p}_i$ would work when the noise is within the bound.

6.2.3   Adaptive Control for ACC

We consider the same ACC case study as in Sect. 4.3 except that we have a time-varying control bound defined as:

$$-c_d(t)Mg \leq u(t) \leq c_a(t)Mg, \forall t \in [0, t_f], \tag{6.23}$$

where $c_d(t) > 0$ and $c_a(t) > 0$ are deceleration and acceleration coefficients, respectively, and g is the gravity constant.

The parameters are $v(0) = 20\,\text{m/s}$, $z(0) = 100\,\text{m}$, $v_p = 13.89\,\text{m/s}$, $v_d = 24\,\text{m/s}$, $M = 1650\,\text{kg}$, $g = 9.81\,\text{m/s}^2$, $f_0 = 0.1\,\text{N}$, $f_1 = 5\,\text{Ns/m}$, $f_2 = 0.25\,\text{Ns}^2/\text{m}$, $l_p = 10\,\text{m}$, $v_{max} = 30\,\text{m/s}$, $v_{min} = 0\,\text{m/s}$, $\Delta t = 0.1\,\text{s}$, $\epsilon = 10$, $c_a(t) = 0.4$. If we apply the HOCBF approach to implement the safety constraint (2.5) with *fixed* $p_1(t) = 0.1$, $p_2(t) = 1$ and $c_d(t) = 0.4$, the QPs will be infeasible after the corresponding HOCBF constraint becomes active. Therefore, we need an AdaCBF to implement this safety constraint, as shown next.

Motivated by (6.22), we first normalize each term in the cost (6.21) by dividing it by its maximum value using: $u_{\lim} = \frac{(c_a Mg)^2}{M^2} = (c_a g)^2$, $v_{1,\max} = \delta_{1,\max}^2 = \delta_{acc,\max}^2 = p_{1,\max}^2 = u_{\lim}$, respectively, where $\delta_{acc,\max}$ denotes the maximum value of δ_{acc}, and set $p_{acc} = c_0 = e^{-12}$, $W_1 = 2e^{-12}$, $P_1 = Q = 0.5$.

Adaptivity to the changing control bound $-c_d(t)Mg$: We first study what happens when we change the lower control bound $-c_d(t)Mg$. In each simulated trajectory, we set the lower control bound coefficient $c_d(t)$ to a different constant or to be time-varying (e.g., linearly decreasing $c_d(t)$). In this case, we set $T = 30s$, $p_1(0) = p_1^* = 0.1$, $p_2^* = 1$. We first present a case of linearly decreasing $c_d(t)$ representing, for example, tires slipping, as shown in Fig. 6.1. When we decrease $c_d(t)$ (weaken the braking capability of the vehicle) after the HOCBF constraint becomes active, the QPs can easily become infeasible in the HOCBF method, as the red curve shows in Fig. 6.1: reading the red curve from right to left as the value of $b(x)$ becomes small as time goes by, the speed is initially stabilized to v_d until the HOCBF constraint becomes active. It is important to note that at some point a QP becomes infeasible. However, using the PACBF method, the QPs are always feasible (blue curve in Fig. 6.1), demonstrating the adaptivity of the PACBF to the time-varying control bound (wheels slipping).

The computational time of the QP at each time step for both the HOCBF and PACBF methods is less than 0.01 s. Note that there is a control overshot when $b(x)$ is small; this can be alleviated by increasing the weight on the control $u(t)$ or by decreasing the weights P_1, Q after the control constraint becomes inactive, as seen by the blue curve in Fig. 6.2.

The simulated trajectories for different (constant) $c_d(t)$ values (e.g., as the vehicle encounters *different road surfaces*) are shown in Figs. 6.2 and 6.3. As seen in Fig. 6.2, when $c_d(t) = 0.4$, the QPs exhibit good feasibility. This induces only a small change in the penalty

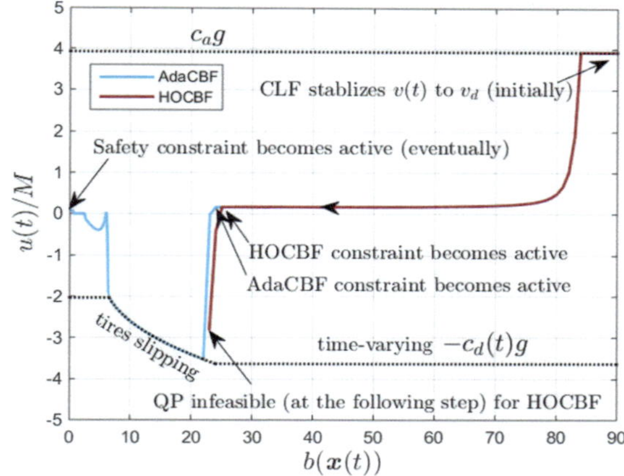

Fig. 6.1 Control input $u(t)$ variation as $b(x(t)) \to 0$ for HOCBF and PACBF for linearly decreasing $c_d(t)$ (0.37 $\to$ 0.2) after the PACBF (or HOCBF) constraint becomes active. The arrow denotes the changing trend for $b(x(t))$ that captures the safety constraint (2.5) (distance between vehicles) with respect to time

Fig. 6.2 Control input $u(t)$ variations as $b(\boldsymbol{x}(t)) \to 0$ under different and time-varying control lower bounds. The arrow denotes the changing trend for $b(\boldsymbol{x}(t))$ with respect to time. $b(\boldsymbol{x}) \geq 0$ implies the *forward invariance* of $C_1 := \{\boldsymbol{x} : b(\boldsymbol{x}) \geq 0\}$

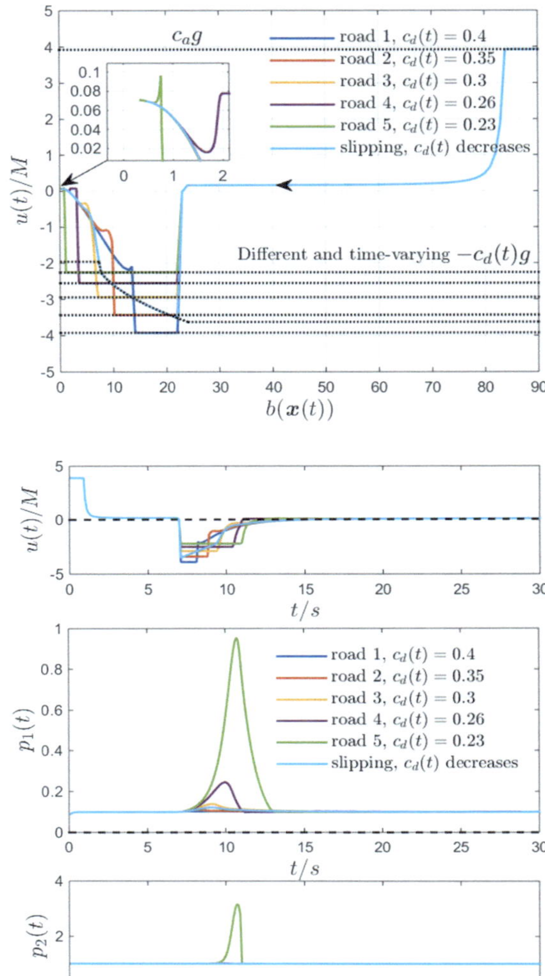

Fig. 6.3 Penalty functions $p_1(t)$, $p_2(t)$ and control input $u(t)$ profiles under different and time-varying control lower bounds. The change in the values of the penalty functions $p_1(t)$, $p_2(t)$ demonstrates the adaptivity of the PACBF to changes in the control bound (or tight control bound)

variable $p_1(t)$ and no change in $p_2(t)$, as shown in Fig. 6.3. As we decrease $c_d(t)$ (i.e., limit the braking capability of the vehicle), the variation in the penalty $p_1(t)$ becomes large after the PACBF constraint becomes active. When $c_d(t) = 0.23$, the vehicle needs to brake with $u(t) = -c_d(t)Mg$ almost all the way to the safety constraint (2.5) becoming active, as the green curves show in Figs. 6.2 and 6.3. On the other hand, the penalty functions $p_1(t)$, $p_2(t)$ both change to a large value, as shown in Fig. 6.3. If we further decrease $c_d(t)$, the safety constraint (2.5) will be violated. The change in the values of the penalty functions $p_1(t)$, $p_2(t)$ demonstrates the adaptivity of the PACBF to changes in the control bound. The penalty method [71] has shown that smaller penalties are needed to improve QP feasibility before the HOCBF constraint becomes active, but the PACBF shows that we may actually want to increase the value of the penalties after the PACBF constraint becomes active, as the last plot in Fig. 6.3 demonstrates.

Adaptivity to noisy dynamics: Suppose we add two noise terms $w_1(t)$, $w_2(t)$ to the speed and acceleration in (4.2), respectively, where $w_1(t)$, $w_2(t)$ denote two random processes defined in an appropriate probability space. In the simulated system, $w_1(t)$, $w_2(t)$ randomly take values in $[-2, 2\,\text{m/s}]$ and $[-0.45, 0.45\,\text{m/s}^2]$ with equal probability at time t, respectively. We fix the value of $c_d(t)$ to 0.23 in the control bound and set $T = 30\,\text{s}$, $p_1(0) = p_1^* = 0.1$, $p_2^* = 1$. The simulation results under different noise levels are shown in Figs. 6.4 and 6.5, the noise is based on $[-2, 2\,\text{m/s}]$ and $[-0.45, 0.45\,\text{m/s}^2]$ for $w_1(t)$, $w_2(t)$, respectively.

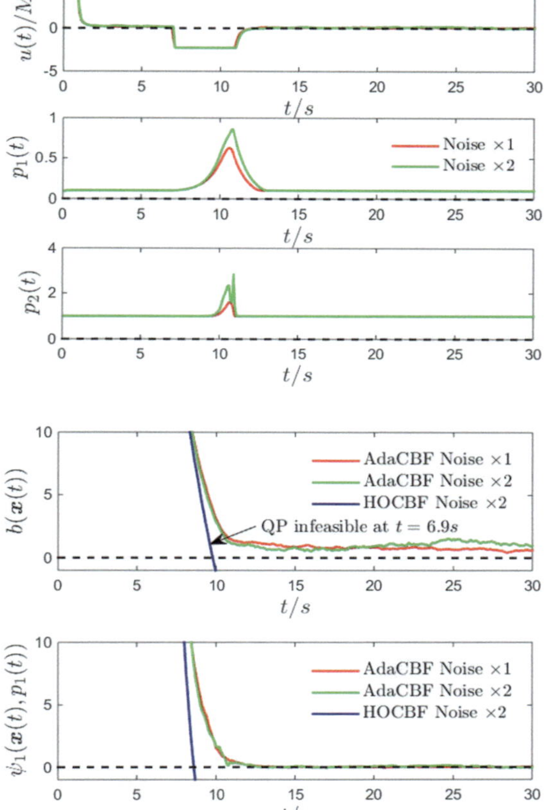

Fig. 6.4 Penalty functions $p_1(t)$, $p_2(t)$ and control input $u(t)$ profiles under different noise levels. The change in the values of the penalty functions $p_1(t)$, $p_2(t)$ demonstrates the adaptivity of the PACBF to the control bound and noise

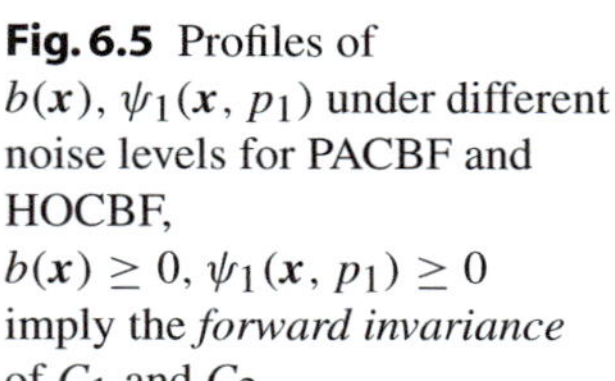

Fig. 6.5 Profiles of $b(\boldsymbol{x})$, $\psi_1(\boldsymbol{x}, p_1)$ under different noise levels for PACBF and HOCBF, $b(\boldsymbol{x}) \geq 0$, $\psi_1(\boldsymbol{x}, p_1) \geq 0$ imply the *forward invariance* of C_1 and C_2

When the control constraint is active, i.e., $u(t) = -c_d(t)Mg$, it can easily conflict with the HOCBF constraint (if we apply the HOCBF method) that is subjected to noise which may make the safety constraint (2.5) violated as the blue line shows in Fig. 6.5. However, the PACBF constraint is relaxed by the penalty functions $p_1(t)$ (through $v_1(t)$) and $p_2(t)$, hence, it is adaptive to different noise levels and can enable the feasibility of the QPs, as seen in Fig. 6.4.

The forward invariance of $C_1 := \{x : b(x) \geq 0\}$ and $C_2 = \{(x, p_1) : \psi_1(x, p_1) \geq 0\}$ is illustrated in Fig. 6.5. Note that $\psi_1(x, p_1)$ might be temporarily negative due to noise during simulation, but will be positive again soon after. This is due to the definition of $\psi_2 := \dot{\psi}_1 + p_2(t)\psi_1$ in (8.9), in which $\alpha_2(\cdot)$ is defined as an extended class $\mathcal{K}$ function (linear function), as shown in [4]. When $\psi_1 < 0$, the PACBF constraint ensures $\dot{\psi}_1 + p_2(t)\psi_1 \geq 0$, thus, $\dot{\psi}_1 \geq -p_2(t)\psi_1 > 0$ since $p_2(t) > 0$. Therefore, ψ_1 will be increasing and eventually becomes positive. In this paper, we have considered high-order polynomial class $\mathcal{K}$ functions to make ψ_i stay away from zero [71] such that $b(x) \geq 0$ is guaranteed in the presence of noise. The forward invariance guarantee can also be achieved by considering the noise bounds in the PACBF constraint. Note that we can also define $\alpha_2(\cdot)$ as a quadratic function in the definition of the PACBF in (8.9) to make $\psi_1(x, p_1)$ also stay away from 0 in Fig. 6.5, and define $\alpha_1(\cdot)$ as a higher-order polynomial function to make the PACBF $b(x)$ stay further away to 0, so that it can be adaptive (in the sense of both QP feasibility and forward invariance) to higher noise levels.

6.3   Relaxation-Adaptive Control Barrier Functions

In this section, we introduce the Relaxation-Adaptive CBF (RACBF). The RACBF works similarly as the PACBF, but aims at achieving adaptivity through a relaxation variable applied to the original constraint instead of introducing penalty functions to the HOCBF in Definition 3.3.

Recall that in PACBFs we define $\psi_0(x) = b(x)$ for a relative degree m function $b(x)$, where $b : \mathbb{R}^n \to \mathbb{R}$, and introduce multiplicative penalty functions to all class $\mathcal{K}$ functions in (6.6) to obtain adaptivity. As an alternative, we may relax ψ_0 as follows:

$$\psi_0(x, r(t)) := b(x) - r(t), \tag{6.24}$$

where $r(t) \geq 0$ is a *relaxation variable* that plays a similar role as penalty functions in a PACBF to obtain adaptivity.

We require that $r(t) \geq 0, \forall t \geq 0$, therefore, we define $r(t)$ to be a HOCBF, similar to the definition of $b(x) \geq 0$ in Definition 3.3. Just like $b(x)$ is associated with the dynamic system (2.11), we need to introduce an *auxiliary dynamic system* for $r(t)$. Moreover, as in Definition 3.3, the relaxation $r(t)$ will be differentiated m times. Thus, we define $R(t) := (r_1(t), r_2(t), \ldots, r_m(t)) \in \mathbb{R}^m$, where $r_1(t) \equiv r(t)$ and $r_j \in \mathbb{R}, j \in \{2, \ldots, m\}$ are the auxiliary state variables for which we define input-output linearizable auxiliary dynamics for $r(t)$ (we henceforth omit the time variable t for simplicity) in the form:

$$\begin{aligned}
\dot{\boldsymbol{R}} &= f_0(\boldsymbol{R}) + g_0(\boldsymbol{R})v, \\
y &= r,
\end{aligned} \tag{6.25}$$

where y denotes the output, $f_0 : \mathbb{R}^m \to \mathbb{R}^m$, $g_0 : \mathbb{R}^m \to \mathbb{R}^m$, and $v \in \mathbb{R}$ denotes the control input for the auxiliary dynamics (6.25). The exact forms of f_0, g_0 may be defined based on a specific application. For simplicity, we usually adopt a linear form. We can initialize $\boldsymbol{r}(0)$ to any real number vector as long as $r(0) > 0$.

In order to properly define the RACBF, we augment system (2.11) with the auxiliary dynamics (6.25) in the form:

$$\begin{bmatrix} \dot{\boldsymbol{x}} \\ \dot{\boldsymbol{R}} \end{bmatrix} = \underbrace{\begin{bmatrix} f(\boldsymbol{x}) \\ f_0(\boldsymbol{R}) \end{bmatrix}}_{F(\boldsymbol{x},\boldsymbol{R})} + \underbrace{\begin{bmatrix} g(\boldsymbol{x}) & \boldsymbol{0} \\ \boldsymbol{0} & g_0(\boldsymbol{R}) \end{bmatrix}}_{G(\boldsymbol{x},\boldsymbol{R})} \begin{bmatrix} \boldsymbol{u} \\ v \end{bmatrix}, \tag{6.26}$$

where $F : \mathbb{R}^{n+m} \to \mathbb{R}^{n+m}$, $G : \mathbb{R}^{n+m} \to \mathbb{R}^{(n+m)\times(q+1)}$. Since r is a HOCBF with relative degree m for (6.25), similar to (3.8), we define a constraint set $U_0(\boldsymbol{R})$ for v:

$$U_0(\boldsymbol{R}) = \{v \in \mathbb{R} : L_{f_0}^m r + [L_{g_0} L_{f_0}^{m-1} r]v + O(r) + \alpha_m(\psi_{m-1}(r)) \geq 0\}, \tag{6.27}$$

where $\psi_{m-1}(r)$ is defined similar to (3.3). With the auxiliary dynamics (6.25), we have $\psi_0(\boldsymbol{x}, \boldsymbol{R}) = b(\boldsymbol{x}) - r$. We define a sequence of functions $\psi_i : \mathbb{R}^{n+m} \to \mathbb{R}, i \in \{1, \ldots, m\}$ in the form:

$$\psi_i(\boldsymbol{x}, \boldsymbol{R}) := \dot{\psi}_{i-1}(\boldsymbol{x}, \boldsymbol{R}) + \alpha_i(\psi_{i-1}(\boldsymbol{x}, \boldsymbol{R})), i \in \{1, \ldots, m\}, \tag{6.28}$$

where $\alpha_i(\cdot), i \in \{1, \ldots, m\}$ denotes a $(m-i)$th order differentiable class $\mathcal{K}$ function. We further define a sequence of sets $C_i, i \in \{1, \ldots, m\}$ associated with (6.28) in the form:

$$C_i := \{(\boldsymbol{x}, \boldsymbol{R}) \in \mathbb{R}^{n+m} : \psi_{i-1}(\boldsymbol{x}, \boldsymbol{R}) \geq 0\}, i \in \{1, \ldots, m\}. \tag{6.29}$$

Definition 6.2 (*Relaxation-Adaptive Control Barrier Function (RACBF)*) Let $C_i, i \in \{1, \ldots, m\}$ be defined by (6.29), $\psi_i(\boldsymbol{x}, \boldsymbol{R}), i \in \{1, \ldots, m\}$ be defined by (6.28) with $\psi_0(\boldsymbol{x}, \boldsymbol{R}) := b(\boldsymbol{x}) - r(t)$, and the auxiliary dynamics be defined by (6.25). A function $b : \mathbb{R}^n \to \mathbb{R}$ is a *Relaxation-Adaptive Control Barrier Function* (RACBF) with relative degree m for (6.26) if r is a HOCBF with relative degree m for the auxiliary dynamics (6.25), and there exist $(m-i)$th, $i \in \{1, \ldots, m-1\}$ order differentiable class $\mathcal{K}$ functions α_i, and a class $\mathcal{K}$ function α_m such that

$$\sup_{u \in U, v \in U_0} [L_F^m \psi_0(x, R) + L_G L_F^{m-1} \psi_0(x, R) \begin{bmatrix} u \\ v \end{bmatrix} + \tag{6.30}$$

$$O(\psi_0(x, R)) + \alpha_m(\psi_{m-1}(x, R))] \geq 0,$$

for all $(x, R) \in C_1 \cap, \ldots, \cap C_m$. In (6.30), $O(\psi_0(x, R))$ denotes the remaining Lie derivative terms of $\psi_0(x, R)$ along F with degree less than m.

Theorem 6.4 ([77]) *Given a RACBF $b(x)$ from Definition 6.2 with the associated sets $C_1, \ldots, C_m$ defined by (6.29), if $(x(0), R(0)) \in C_1 \cap, \ldots, \cap C_m$, then any Lipschitz continuous controller $(u(t), v(t))$ that satisfies (6.30), $\forall t \geq 0$ renders $C_1 \cap, \ldots, \cap C_m$ forward invariant for system (6.26), and renders $b(x(t)) \geq 0, \forall t \geq 0$ for system (2.11).*

Proof It follows from Theorem 3.2 that the set $C_1 \cap, \ldots, \cap C_m$ is forward invariant for system (6.26). Since $r(t)$ is a HOCBF, we also have $r(t) \geq 0, \forall t \geq 0$. Since C_1 is forward invariant, we have $\psi_0(x(t), R(t)) = b(x(t)) - r(t) \geq 0, \forall t \geq 0$. Therefore, $b(x(t)) \geq r(t) \geq 0, \forall t \geq 0$. $\square$

Remark 6.4 (*Adaptivity of RACBF*) The control u of system (2.11) depends on the control v of the auxiliary system (6.25) in the RACBF constraint (6.30). Similar to the PACBF, the control v is only constrained by the HOCBF constraint in (6.27) since we require that r is a HOCBF. Therefore, we also relax the constraints on the control input of system (2.11) in the RACBF by allowing the relaxation r to be time-varying. The satisfaction of the original constraint $b(x) \geq 0$ is still guaranteed. This is how a RACBF provides "adaptivity." This adaptivity can guarantee problem feasibility under time-varying control bounds and noisy dynamics if some additional conditions are satisfied, as will be discussed in the next section.

6.3.1 Optimal Control with RACBFs

Consider an optimal control problem for system (2.11) with the cost defined as (6.16), and the time-varying control bounds defined as (6.17). Assume a (safety) constraint $b(x) \geq 0$ with relative degree m has to be satisfied by system (2.11). We use the RACBF method to guarantee $b(x) \geq 0$ so that u should satisfy the RACBF constraint (6.30). Moreover, each v is constrained by the HOCBF constraint (6.27) corresponding to the constraint $r(t) \geq 0$ for the auxiliary dynamics (6.25).

However, if $R(t) = 0, \forall t \in [0, T]$, we have $v = 0$ following from (6.25) and (6.27). Then, a RACBF loses its adaptivity as the control u is not relaxed by v in the RACBF constraint (6.30). Similar to the penalty function $p_i(t)$ of a PACBF, we also use a CLF to stabilize $r(t)$ to some desirable value $r^* > 0$. Assuming the auxiliary dynamics (5.9) are input-output linearized (otherwise, we perform input-output linearization), we can use either the tracking control from [26] or the CLF to stabilize $r(t)$, i.e., if $m = 1$, we define a CLF $V(r) := (r - r^*)^2$, and if $m \geq 2$, we find a desired state feedback form $\hat{r}_m$ for r_m (note that $r_1 \equiv r$ in (6.25)):

$$\hat{r}_m = \begin{cases} -k_1(r - r^*), & m = 2, \\ -k_1(r-r^*)-k_2r_2-\cdots, -k_{m-1}r_{m-1}, & m > 2, \end{cases} \tag{6.31}$$

where $k_1 > 0, \ldots, k_{m-1} > 0$.

Remark 6.5 The desired value r^* should be chosen such that the problem itself is feasible when the RACBF constraint (6.30) first becomes active. We may find some $\bar{r} > 0$ such that the problem is feasible under the worst-case conditions (e.g., the maximum approaching speed of a vehicle towards another vehicle) and worst-case noise (i.e., its bound). Then any $r^* \geq \bar{r}$ would work when the noise is within the bound. The exact choice of r^* depends on the particular application. For the adaptive cruise control problem, we should choose r^* to be no less than the minimum braking distance as in [2].

We can now define a CLF $V(\boldsymbol{R}) := (r_m - \hat{r}_m)^2$ (the relative degree of $V(\boldsymbol{R})$ is one) to stabilize r so that any control input v should satisfy:

$$L_{f_0}V(\boldsymbol{R}) + L_{g_0}V(\boldsymbol{R})v + \epsilon V(\boldsymbol{R}) \leq \delta_r, \tag{6.32}$$

where δ_r is a relaxation variable that we want to minimize. Therefore, we can reformulate the cost (6.16) to incorporate the RACBF as follows:

$$\min_{\boldsymbol{u}(t),v(t),\delta_r(t)} \int_0^T [\mathcal{C}(||\boldsymbol{u}(t)||)+P_r v^2(t) + P_r \delta_r^2(t)]dt \tag{6.33}$$

subject to (2.11), (6.17), (6.30), (6.25), the HOCBF constraint in (6.27) for $r \geq 0$, and the CLF constraint (6.32). In (6.33), $P_r > 0$. We can then use the QP-based approach introduced at the end of Chap. 2 to solve (6.33). We can also normalize each term in (6.33) and form a convex combination as in (6.22).

In (6.33), the time complexity is still polynomial in the dimension d of the decision variables, where $d = q + 2$. Note that this is smaller than $q + 2m - 1$ in the PACBF method (the relative degree m does not affect the complexity).

Remark 6.6 (*Forward invariance of RACBF with noise*) In order to make sure that $b(\boldsymbol{x}) \geq 0$ is guaranteed under noisy dynamics, we can make $r \geq r_a > 0$ instead of $r \geq 0$, i.e., define $r - r_a$ as a HOCBF. The value of r_a depends on the magnitude of the noise level.

6.3.2 Comparison Between PACBF and RACBF

The PACBF achieves adaptivity by using penalty functions in its definition and this can alleviate the conservativeness of the CBF method. On the other hand, the RACBF achieves adaptivity through a relaxation variable in the original constraint. The definition of a RACBF is close to a HOCBF in Definition 3.3 such that conservativeness still exists, while a PACBF is not conservative following Theorem 6.2. Therefore, a PACBF has better adaptivity properties than a RACBF. However, as already stated, RACBF has better time complexity than PACBF. The trade-off between adaptivity and complexity depends on the available computational resources and PACBF is preferable if computational resources are available. We can also combine these two alternative approaches by simultaneously adding a relaxation variable as in (6.24) and multiplying (partially) by penalty functions as in (6.6).

6.3.3 Control for ACC Using RACBF

We use the same problem setup as in Sect. 6.2.3. We first normalize each term in the cost (6.33) by dividing its maximum value: $u_{\lim} = v_{\max}^2 = \delta_{r,\max}^2 = \delta_{acc,\max}^2 = (c_a g)^2$, respectively, where $v_{\max}$ denotes the maximum value of v, and set $c_0 = p_{acc} = e^{-4}$, $P_r = 0.5$, $r_a = 1m$, $r(0) = (3, 0)$.

We also change the lower control bound $-c_d(t)Mg$. In each simulated trajectory, we set the lower control bound coefficient $c_d(t)$ to a different constant or to be time-varying (e.g., linearly decreasing $c_d(t)$). In this case, we set $T = 30s$, $p_1 = 0.1$, $p_2 = 1$. We first present a case of linearly decreasing $c_d(t)$ representing, for example, tires slipping, as shown in Fig. 6.6. When we decrease $c_d(t)$ (weaken the braking capability of the vehicle) after the HOCBF constraint becomes active, the QPs can easily become infeasible in the HOCBF

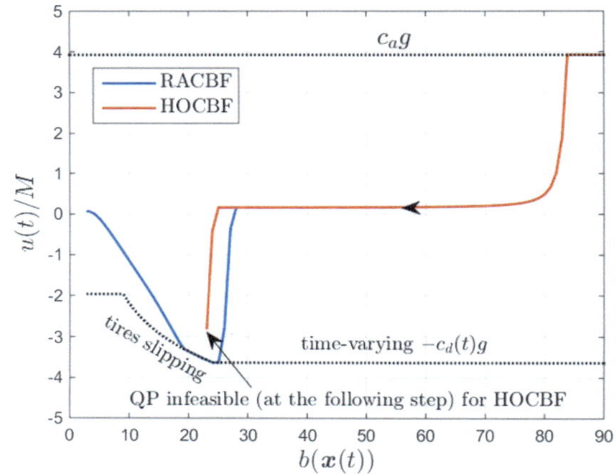

Fig. 6.6 Control input $u(t)$ variation as $b(\boldsymbol{x}(t)) \to 0$ for HOCBF and RACBF for linearly decreasing $c_d(t)$ ($0.37 \to 0.2$) after the RACBF (or HOCBF) constraint becomes active

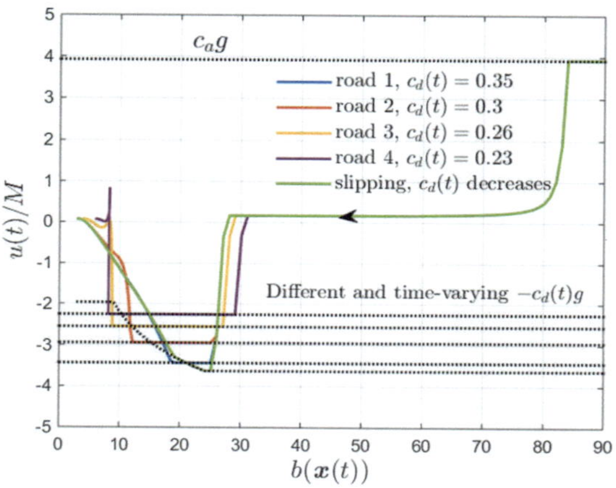

Fig. 6.7 Control input $u(t)$ variations as $b(\boldsymbol{x}(t)) \to 0$ under different and time-varying control lower bounds for RACBFs

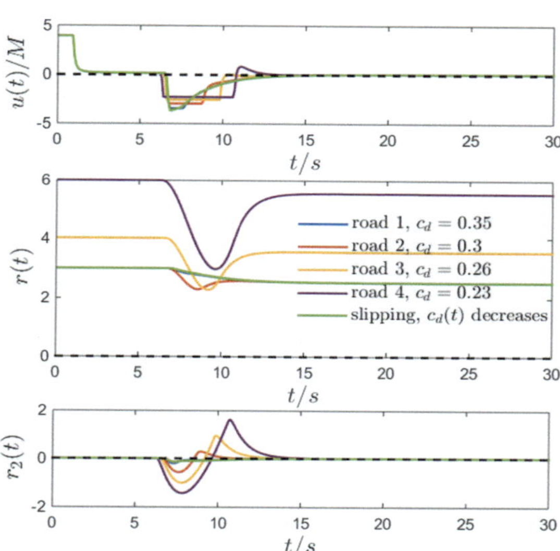

Fig. 6.8 Relaxation $r(t)$, $r_2(t)$ and control input $u(t)$ profiles under different and time-varying control lower bounds. The change in the values of the relaxation $r(t)$, $r_2(t)$ demonstrates the adaptivity of the RACBF to changes in the control bound (or tight control bound)

method, as the brown curve shows in Fig. 6.6. It is important to note that at some point a QP becomes infeasible. However, using the RACBF method, the QPs are always feasible (blue curve in Fig. 6.6), demonstrating the adaptivity of the RACBF to the time-varying control bound (wheels slipping).

The computational time at each time step for the RACBF method is less than $0.01\,$s in MATLAB (*Intel(R) Core(TM) i7-8700 CPU @ 3.2 GHz*×2). The simulated trajectories for different (constant) $c_d(t)$ values (e.g., as the vehicle encounters *different road surfaces*) are shown in Figs. 6.7 and 6.8. Similar to the PACBF, the RACBF can also be adaptive to the time-varying control bounds. The adaptivity of RACBF to noise is shown in what follows

Comparison between PACBF and RACBF. Here, we compare the PACBF and RACBF methods. We set $c_0 = p_{acc} = 0.005$, $P_r = 0.495$, $r_a = 1m$ for (6.33) and set $p_{acc} = c_0 = e^{-12}$, $W_1 = 2e^{-12}$, $P_1 = Q = 0.5$ for (6.22). The lower control bound coefficient is $c_d(t) = 0.23$. $p_1^* = 0.1$, $p_2^* = 1$ for the PACBF, and $k_1 = 0.1$, $k_2 = 1$ for the RACBF. In the simulated system, $w_1(t)$, $w_2(t)$ randomly take values in $[-4, 4\,\text{m/s}]$ and $[-0.9, 0.9\text{m/s}^2]$ with equal probability at time t, respectively. The simulation results are shown in Figs. 6.9 and 6.10.

It follows from Fig. 6.9 that the RACBF method is still conservative such that it cannot utilize the space marked in the figure to obtain adaptivity. This is due to the fact that the RACBF method achieves adaptivity by the relaxation $r(t)$, and this will make the RACBF constraint become active *earlier* than the PACBF method, as shown in Fig. 6.9. The PACBF is not conservative, and thus is more adaptive than the RACBF. However, the complexity of a PACBF-based method increases faster than the one of a RACBF-based approach as the

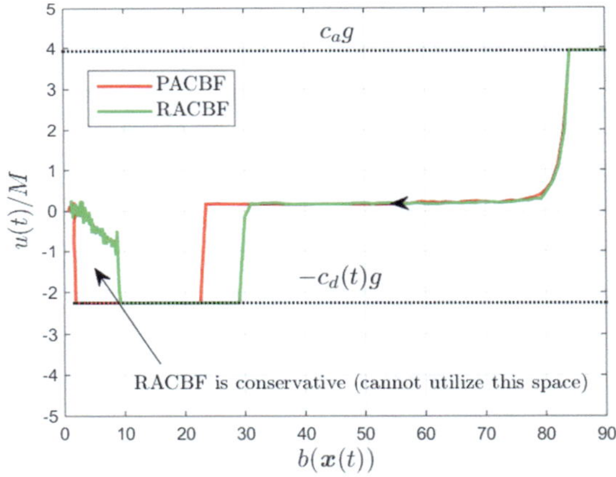

Fig. 6.9 Adaptivity comparison between the PACBF and RACBF methods

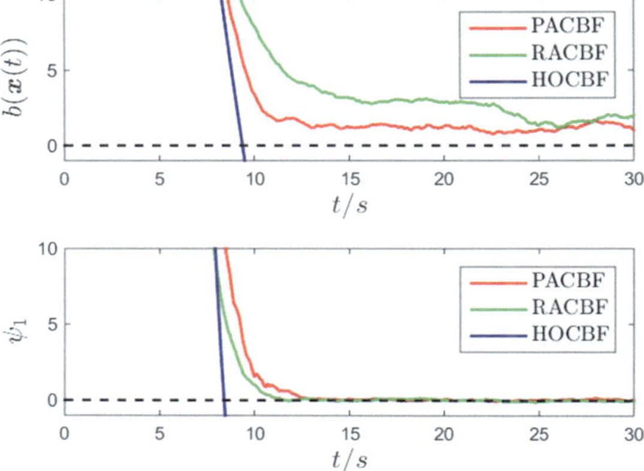

Fig. 6.10 Profiles of $b(x)$, $\psi_1(x, p_1)$ under noise for PACBF, RACBF and HOCBF, $b(x) \geq 0$ implies the *forward invariance* of $C_1 = \{x : b(x) \geq 0\}$

relative degree m of the constraint becomes larger. It follows from Fig. 6.10 that the satisfaction of $b(x) \geq 0$ under noise is guaranteed with both the PACBF and RACBF methods, thus the QPs used in solving the associated optimal control problem are always feasible.

Safety Guarantees for Systems with Unknown Dynamics: An Event-Driven Framework

7

7.1 Adaptive Affine Dynamics

We consider a system with state $x \in X$ and control $u \in U$ with unknown dynamics, as shown in Fig. 7.1. For the unknown dynamics, we make the following assumption:

Assumption 7.1 The relative degree of each component of x is known with respect to the real unknown dynamics.[1]

For example, if the position of a vehicle (whose dynamics are unknown) is a component in x and the control is its acceleration, then the relative degree of the position with respect to the unknown vehicle dynamics is two by Newton's law. We assume that our system is also equipped with sensors capable of monitoring x and its derivatives.

We now return to problem **OCP** introduced in Sect. 2.5, i.e., we wish to find a control policy that satisfies safety and control constraints (2.18) and (2.19) for system (2.11) and such that the cost (2.17) is minimized. When the dynamics are unknown, a control policy is *feasible* if constraints (2.18) and (2.19) are satisfied at all times. We then consider the following problem **OCP-U**, which differs from **OCP** in that the real system dynamics are now unknown:

OCP-U: Find a feasible control policy for the real unknown dynamics such that the cost (2.17) is minimized. $\qquad\qquad\square$

---

[1] The relative degree is as defined in Definition 3.1 for the real unknown dynamics.

© The Author(s), under exclusive license to Springer Nature Switzerland AG 2023
W. Xiao et al., *Safe Autonomy with Control Barrier Functions*,
Synthesis Lectures on Computer Science,
https://doi.org/10.1007/978-3-031-27576-0_7

Our approach to solve **OCP-U** relies on the CBF-based QP method (2.21). There are four steps involved in the solution:

Step 1: Define adaptive affine dynamics. Our motivation is that we need affine dynamics of the form (2.11) in order to apply the CBF-based QP approach to solve **OCP-U**. Under Assumption 7.1, we define affine dynamics that have the same relative degree for (2.18) as the real system to estimate the real unknown dynamics in the form:

$$\dot{\bar{x}} = f_a(\bar{x}) + g_a(\bar{x})u, \tag{7.1}$$

where $f_a : \mathbb{R}^n \to \mathbb{R}$, $g_a : \mathbb{R}^n \to \mathbb{R}^{n \times q}$, and $\bar{x} \in X \subset \mathbb{R}^n$ is the state vector corresponding to x in the unknown dynamics. Since $f_a(\cdot)$, $g_a(\cdot)$ in (7.1) can be adaptively updated to accommodate the real unknown dynamics, as shown in the next section, we refer to (7.1) as *adaptive affine dynamics*. The real unknown dynamics and (7.1) are related through the error states obtained from the real-time measurements of the system and the integration of (7.1). Theoretically, we can take any affine dynamics in (7.1) to model the real system as long as their states are of the same dimension and with the same physical interpretation within the plant. Clearly, however, we would like the adaptive dynamics (7.1) to "stay close" to the real dynamics. This notion will be formalized in the next section.

Step 2: Find a HOCBF that guarantees the safety constraints (2.18). Based on (7.1), the error state and its derivatives, we use a HOCBF to enforce (2.18). Details are shown in the next section.

Step 3: Formulate the CBF-based QPs. We use a relaxed CLF to achieve a minimal value of the terminal state penalty in (2.17). If $C(\|u(t)\|) = \|u(t)\|^2$ in (2.17), then we can formulate problem **OCP-U** using the same CBF-CLF-QP approach as in (2.21), with a CBF replaced by a HOCBF if $m > 1$ as described in Chap. 3.

Step 4: Determine the events required to solve each QP and the condition that guarantees the satisfaction of (2.18) **between events**. Since there is obviously a difference between the adaptive affine dynamics (7.1) and the real unknown dynamics, in order to guarantee safety in the real system, we need to properly define *events* which are dependent on the error state and the state of (7.1) in order to solve the QP. In other words, we need to determine the times t_k, $k = 1, 2, \ldots$ $(t_1 = 0)$ at which the QP must be solved in order to guarantee the satisfaction of (2.18) for the real unknown dynamics.

The overall solution framework is outlined in Fig. 7.1 where we note that we apply the same control from each QP solution to both the real unknown dynamics and (7.1).

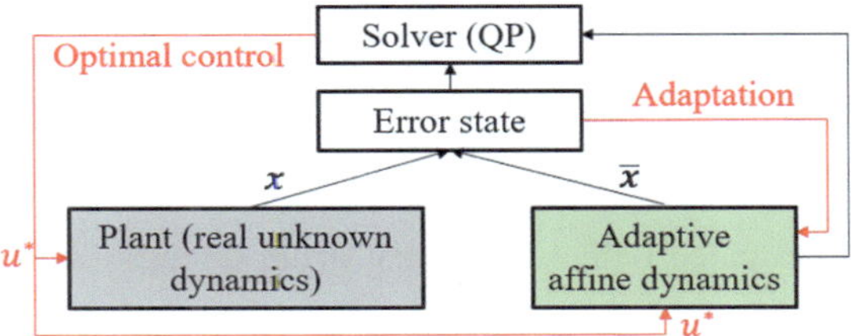

Fig. 7.1 Graphical illustration of our approach to **OCP-U** and the connection between the real unknown dynamics and the adaptive affine dynamics (7.1). The state x is obtained from sensor measurements of the plant

7.2   Event-Driven Control

In this section, we provide the details involved in formulating the CBF-based QPs that guarantee the satisfaction of the safety constraints (2.18) for the real unknown system. We begin with the case of relative-degree-one safety constraints (2.18).

7.2.1   Relative-Degree-One Constraints

Suppose the safety constraint in (2.18) has relative degree one with respect to both dynamics (7.1) and the actual dynamics (2.11). Next, we show how to find a CBF that guarantees (2.18) for the real unknown dynamics. Let

$$e := x - \bar{x} \tag{7.2}$$

and note that x and $\bar{x}$ are state vectors from direct measurements and from the adaptive dynamics (7.1), respectively. Then,

$$b(x) = b(\bar{x} + e). \tag{7.3}$$

Differentiating $b(\bar{x} + e)$, we have

$$\frac{db(\bar{x} + e)}{dt} = \frac{\partial b(\bar{x} + e)}{\partial \bar{x}} \dot{\bar{x}} + \frac{\partial b(\bar{x} + e)}{\partial e} \dot{e}. \tag{7.4}$$

The CBF constraint that guarantees (2.18) for known dynamics (2.11) is as defined in (2.13), where we replace $\dot{x}$ with (2.11). However, for the unknown dynamics, the CBF constraint is $\frac{db(x)}{dt} + \alpha_1(b(x)) \geq 0$. Equivalently, we have

$$\frac{db(\bar{x} + e)}{dt} + \alpha_1(b(\bar{x} + e)) \geq 0. \tag{7.5}$$

Combining (7.4), (7.5) and (7.1), we get the CBF constraint that guarantees (2.18):

$$\frac{\partial b(x)}{\partial \bar{x}} f_a(\bar{x}) + \frac{\partial b(x)}{\partial \bar{x}} g_a(\bar{x}) u + \frac{\partial b(x)}{\partial e} \dot{e} + \alpha_1(b(x)) \geq 0. \tag{7.6}$$

where $\dot{e} = \dot{x} - \dot{\bar{x}}$ is evaluated online through $\dot{x}$ (from direct measurements of the actual state derivative) and $\dot{\bar{x}}$ is given through (7.1). Then, the satisfaction of (7.6) implies the satisfaction of $b(\bar{x} + e) \geq 0$ by Theorem 3.2 and (7.3). Therefore, (2.18) is guaranteed to be satisfied for the real unknown dynamics.

Now, we can formulate a CBF-based optimal control problem in the form:

$$\min_{u(t),\delta(t)} \int_0^T [||u(t)||^2 + p\delta^2(t)]dt \tag{7.7}$$

subject to (7.6), (2.19), and the CLF constraint

$$L_{f_a} V(\bar{x}) + L_{g_a} V(\bar{x}) u + \epsilon V(\bar{x}) \leq \delta(t), \tag{7.8}$$

where $V(\bar{x}) = ||\bar{x} - K||^2$, $p > 0$, and $\delta(t)$ is a relaxation variable for the CLF constraint. Following the approach introduced in Sect. 2.5.1, we solve this problem by discretizing time and solving a sequence of QPs of the form (2.21) at each time step that starts at $t_k, k = 1, 2 \ldots$. However, at time t_k, the associated QP does not generally know the error state $e(t)$ and its derivative $\dot{e}(t)$, $\forall t > t_k$. Thus, it cannot guarantee that the CBF constraint (7.6) is satisfied in the time interval $(t_k, t_{k+1}]$, where t_{k+1} is the next time instant when a QP is solved.

In order to find a condition that guarantees the satisfaction of (7.6) $\forall t \in (t_k, t_{k+1}]$, we first let $e = (e_1, \ldots, e_n)$ and $\dot{e} = (\dot{e}_1, \ldots, \dot{e}_n)$ be bounded by $w = (w_1, \ldots, w_n) \in \mathbb{R}^n_{>0}$ and $v = (v_1, \ldots, v_n) \in \mathbb{R}^n_{>0}$:

$$|e_i| \leq w_i, \qquad |\dot{e}_i| \leq v_i, \qquad i \in \{1, \ldots, n\}. \tag{7.9}$$

These two inequalities can be rewritten in the form $|e| \leq w$, $|\dot{e}| \leq v$ for notational simplicity.

We now consider any state $\bar{x}$ at time t_k, which satisfies:

$$\bar{x}(t_k) - s \leq \bar{x} \leq \bar{x}(t_k) + s, \tag{7.10}$$

where the inequalities are interpreted componentwise and $s \in \mathbb{R}^n_{>0}$. The choice of the parameter vector s will be discussed later. We denote the set of states that satisfy (7.10) at time t_k by

$$S(t_k) = \{y \in X : \bar{x}(t_k) - s \leq y \leq \bar{x}(t_k) + s\}. \tag{7.11}$$

Now, with (7.9) and (7.10), we are ready to find a condition that guarantees the satisfaction of (7.6) in the time interval $(t_k, t_{k+1}]$. This is done by considering the minimum value of each component in (7.6), as shown next.

In (7.6), let $b_{f_a}^{min}(t_k) \in \mathbb{R}$ be the minimum value of $\frac{\partial b(\bar{x}+e)}{\partial \bar{x}} f_a(\bar{x})$ for the preceding time interval that satisfies $y \in S(t_k)$, $|e| \leq w$, $y + e \in C_1$ starting at time t_k, i.e., let

$$b_{f_a}^{min}(t_k) = \min_{y \in S(t_k),|e| \leq w, y+e \in C_1} \frac{\partial b(y + e)}{\partial y} f_a(y) \tag{7.12}$$

Similarly, we can also find the minimum value $b_{\alpha_1}^{min}(t_k) \in \mathbb{R}$ and $b_e^{min}(t_k) \in \mathbb{R}$ of $\alpha_1(b(x))$ and $\frac{\partial b(x)}{\partial e} \dot{e}$, respectively, for the preceding time interval that satisfies $y \in S(t_k)$, $|e| \leq w$, $y + e \in C_1$, $|\dot{e}| \leq v$ starting at time t_k, i.e., let

$$b_{\alpha_1}^{min}(t_k) = \min_{y \in S(t_k),|e| \leq w, y+e \in C_1} \alpha_1(b(y + e)) \tag{7.13}$$

$$b_e^{min}(t_k) = \min_{y \in S(t_k),|e| \leq w,|\dot{e}| \leq v, y+e \in C_1} \frac{\partial b(y + e)}{\partial e} \dot{e} \tag{7.14}$$

This leaves one more term to consider in (7.6): $\frac{\partial b(x)}{\partial \bar{x}} g_a(\bar{x})$. If this term is independent of $\bar{x}$ and e, then we do not need to find its limit value within the set of y bounds $S(t_k)$ and $|e| \leq w$, $y + e \in C_1$; otherwise, let $\bar{x} = (\bar{x}_1, \ldots, \bar{x}_n) \in \mathbb{R}^n$, $u = (u_1, \ldots, u_q) \in \mathbb{R}^q$ and $g_a = (g_1, \ldots, g_q) \in \mathbb{R}^{n \times q}$ and proceed as follows.

We assume each component of $\frac{\partial b(x)}{\partial \bar{x}} g_a(\bar{x})$ does not change sign $\forall x \in X$; otherwise, we can define each sign change to be an additional update-triggering event. The sign of $u_i(t_k)$, $i \in \{1, \ldots, q\}$, $k = 1, 2 \ldots$ can be determined by solving each CBF-based QP associated with (7.7) at t_k. Then, we can determine the limit value $b_{g_i}^{lim}(t_k) \in \mathbb{R}$, $i \in \{1, \ldots, q\}$ of $\frac{\partial b(x)}{\partial \bar{x}} g_i(\bar{x})$, by

$$b_{g_i}^{lim}(t_k) = \begin{cases} \displaystyle\min_{y \in S(t_k),|e| \leq w, y+e \in C_1} \frac{\partial b(y+e)}{\partial y} g_i(y), & \text{if } u_i(t_k) \geq 0, \\[2ex] \displaystyle\max_{y \in S(t_k),|e| \leq w, y+e \in C_1} \frac{\partial b(y+e)}{\partial y} g_i(y), & \text{otherwise} \end{cases} \tag{7.15}$$

Let $b_{g_a}^{lim}(t_k) = (b_{g_1}^{lim}(t_k), \ldots, b_{g_q}^{lim}(t_k)) \in \mathbb{R}^q$, and we set $b_{g_a}^{lim}(t_k) = \frac{\partial b(x)}{\partial \bar{x}} g_a(\bar{x})$ if $\frac{\partial b(x)}{\partial \bar{x}} g(\bar{x})$ is independent of $\bar{x}$ and e for notational simplicity.

The condition that guarantees the satisfaction of (7.6) in the time interval $(t_k, t_{k+1}]$ is then given by

$$b_{f_a}^{min}(t_k) + b_{g_a}^{lim}(t_k) u(t_k) + b_e^{min}(t_k) + b_{\alpha_1}^{min}(t_k) \geq 0. \tag{7.16}$$

In order to apply the above condition to problem (7.7), we just replace (7.6) by (7.16), i.e., we have

$$\min_{u(t),\delta(t)} \int_0^T [||u(t)||^2 + p\delta^2(t)]dt \tag{7.17}$$

subject to (7.16), (2.19) and (7.8). As in Chap. 2, (7.17) is solved through a sequence of QPs of the form (2.21). However, rather than solving each QP based on a specific time discretization (which requires a prespecified time step value), we define instead three events that determine the triggering conditions for solving any such QP:

- **Event 1**: $|e| \leq w$ is about to be violated.
- **Event 2**: $|\dot{e}| \leq v$ is about to be violated.
- **Event 3**: the state of (7.1) reaches the boundaries of $S(t_k)$.

In other words, the next time instant $t_{k+1}, k = 1, 2 \ldots$ to solve each QP associated with (7.17) is determined by:

$$t_{k+1} = \min\{t > t_k : |e(t)| = w \text{ or } |\dot{e}(t)| = v \text{ or } |\bar{x}(t) - \bar{x}(t_k)| = s\}, \qquad (7.18)$$

where $t_1 = 0$. The first two events can be detected by direct sensor measurements, while Event 3 can be detected by monitoring the dynamics (7.1). The magnitude of each component of s is a trade-off between the time complexity and the conservativeness of this approach. If the magnitude is large, then the number of events is small at the expense of making this approach more conservative as we determine the condition (7.16) through the minimum values as in (7.12)–(7.15).

Formally, we have the following theorem to show that the satisfaction of the safety constraint (2.18) is guaranteed for the real unknown dynamics through condition (7.16):

Theorem 7.1 ([75]) *Given a HOCBF $b(x)$ with $m = 1$ as in Definition 3.3, let $t_{k+1}, k = 1, 2 \ldots$ be determined by (7.18) with $t_1 = 0$, and $b_{f_a}^{min}(t_k), b_{\alpha_1}^{min}(t_k), b_e^{min}(t_k), b_{g_a}^{lim}(t_k)$ be determined by (7.12)–(7.15), respectively. Then, under Assumption 7.1, any control $u(t_k)$ that satisfies (7.16) and updates the real unknown dynamics and the adaptive dynamics (7.1) within time interval $[t_k, t_{k+1})$ renders the set C_1 forward invariant for the real unknown dynamics.*

Proof By (7.18), we have:

$$y(t) \in S(t_k), \ |e(t)| \leq w, \ |\dot{e}(t)| \leq v, \ y(t) + e(t) \in C_1$$

for all $t \in [t_k, t_{k+1}], k = 1, 2 \ldots$. Thus, the limit values $b_{f_a}^{min}(t_k), b_{\alpha_1}^{min}(t_k), b_e^{min}(t_k)$, determined by (7.12), (7.13), (7.14), respectively, are the minimum values for $\frac{\partial b(x)}{\partial \bar{x}} f_a(\bar{x}(t))$, $\alpha_1(b(x(t))), \frac{\partial b(x(t))}{\partial e(t)} \dot{e}(t)$ for all $t \in [t_k, t_{k+1})$. In other words, we have

$$\frac{\partial b(x)}{\partial \bar{x}} f_a(\bar{x}(t)) + \alpha_1(b(x(t))) + \frac{\partial b(x(t))}{\partial e(t)} \dot{e}(t) \geq b_{f_a}^{min}(t_k)$$

$$+ b_{\alpha_1}^{min}(t_k) + b_e^{min}(t_k), \ \forall t \in [t_k, t_{k+1}).$$

where $x = e + \bar{x}$.

By (7.15), we have:

$$\frac{\partial b(\boldsymbol{x})}{\partial \bar{\boldsymbol{x}}} g_a(\bar{\boldsymbol{x}}(t))\boldsymbol{u}(t_k) \geq b_{g_a}^{lim}(t_k)\boldsymbol{u}(t_k), \quad \forall t \in [t_k, t_{k+1}).$$

Following the last two equations and (7.16), we have

$$\frac{\partial b(\boldsymbol{x})}{\partial \bar{\boldsymbol{x}}} f_a(\bar{\boldsymbol{x}}(t)) + \frac{\partial b(\boldsymbol{x})}{\partial \bar{\boldsymbol{x}}} g_a(\bar{\boldsymbol{x}}(t))\boldsymbol{u}(t_k) + \alpha_1(b(\boldsymbol{x}(t)))$$
$$+ \frac{\partial b(\boldsymbol{x}(t))}{\partial \boldsymbol{e}(t)}\dot{\boldsymbol{e}}(t) \geq 0, \quad \forall t \in [t_k, t_{k+1}).$$

By (7.5), (7.6) and the last equation, we obtain

$$\frac{db(\bar{\boldsymbol{x}}(t) + \boldsymbol{e}(t))}{dt} + \alpha_1(b(\bar{\boldsymbol{x}}(t) + \boldsymbol{e}(t))) \geq 0, \quad \forall t \in [t_k, t_{k+1}), \ k = 1, 2 \ldots$$

By Theorem 3.2, we have $b(\bar{\boldsymbol{x}}(t) + \boldsymbol{e}(t)) \geq 0$, $\forall t \geq 0$ and by (7.3) it follows that C_1 is forward invariant for the real unknown dynamics. $\qquad\square$

Remark 7.1 We may also consider the minimum value of $\frac{\partial b(\boldsymbol{y}+\boldsymbol{e})}{\partial \boldsymbol{y}} f_a(\boldsymbol{y}) + \frac{\partial b(\boldsymbol{y}+\boldsymbol{e})}{\partial \boldsymbol{e}}\dot{\boldsymbol{e}} + \alpha_1(b(\boldsymbol{y} + \boldsymbol{e}))$ within the bounds $\boldsymbol{y} \in S(t_k)$, $|\boldsymbol{e}| \leq \boldsymbol{w}$, $\boldsymbol{y} + \boldsymbol{e} \in C_1$, $|\dot{\boldsymbol{e}}| \leq \boldsymbol{v}$ instead of considering them separately as in (7.12)–(7.15). This could be less conservative as the constraint (7.16) is stronger compared with the CBF constraint (7.6), and we wish to find the largest possible value of the left-hand side of (7.6) that can support the proof of Theorem 7.1.

Events 1 and 2 will be frequently triggered if the modelling of the adaptive dynamics (7.1) has a large error with respect to the real dynamics. Therefore, we would like to model the adaptive dynamics (7.1) as accurately as possible in order to reduce the number of events required to solve (7.17) through a sequence of associated QPs.

An additional important step is to *synchronize* the state of the real unknown dynamics and (7.1) such that we always enforce $\boldsymbol{e}(t_k) = 0$ and $\dot{\boldsymbol{e}}(t_k)$ as close to 0 as possible by setting

$$\bar{\boldsymbol{x}}(t_k) = \boldsymbol{x}(t_k), \tag{7.19}$$

and by updating $f_a(\bar{\boldsymbol{x}}(t))$ of the adaptive dynamics (7.1) right after an event occurs at t:

$$f_a(\bar{\boldsymbol{x}}(t^+)) = f_a(\bar{\boldsymbol{x}}(t^-)) + \sum_{i=0}^{k} \dot{\boldsymbol{e}}(t_i). \tag{7.20}$$

where t^+, t^- denote instants right after and right before t. In this way, the dynamics (7.1) are adaptively updated at each event, i.e., at t_k, $k = 1, 2, \ldots$. Note that we may also update $g_a(\cdot)$, a more challenging task than updating $f_a(\cdot)$ since $g_a(\cdot)$ is multiplied by $\boldsymbol{u}$ that is to be determined, i.e., the update of $g_a(\cdot)$ will depend on $\boldsymbol{u}$.

Enforcing (7.19) and (7.20) ensures that $e(t_k) = 0$ and $\dot{e}(t_k)$ is close to 0. It is also easy to check that there is no possibility of any Zeno behavior [26]. In particular, there exist lower bounds for the occurrence times of Events 1 and 3 which are determined by the limit values of the component of f_a, g_a within X and U, as well as the real unknown dynamics (although they are unknown). Assuming the functions that define the real unknown dynamics are Lipschitz continuous, and the functions f_a, g_a in (7.1) are also assumed to be Lipschitz continuous, it follows that $\dot{e}$ is also Lipschitz continuous. Suppose the largest Lipschitz constant among all the components in $\dot{x}$ is L, $\forall x \in X$, and the smallest Lipschitz constant among all the components in $\dot{\bar{x}}$ is $\bar{L}$, $\forall \bar{x} \in X$. Then, the lower bound time for Event 2 is $\frac{v_{min}}{L - \bar{L}}$, where $v_{min} > 0$ is the minimum component in v. On the other hand, since the control is constant in each time interval and there exists a lower bound for the event time, Zeno behavior will not occur.

We summarize the event-driven control scheme in Algorithm 1.

Algorithm 1: Event-driven control

Input: Measurements x and $\dot{x}$ from a plant, adaptive model (7.1), settings for QP (7.17), w, v, s.
Output: Event time t_k, $k = 1, 2, \ldots$ and $u^*(t_k)$.
$k = 1, t_k = 0$;
while $t_k \leq T$ **do**
    Measure x and $\dot{x}$ from the plant at t_k;
    Synchronize the state of (7.1) and the plant by (7.19), (7.20);
    Evaluate (7.12)–(7.15);
    Solve the QP associated to (7.17) at t_k and get $u^*(t_k)$;
    while $t \leq T$ **do**
        Apply $u^*(t_k)$ to the plant and (7.1) for $t \geq t_k$;
        Measure x and $\dot{x}$ from the plant;
        Evaluate t_{k+1} by (7.18);
        if t_{k+1} *is found with* $\varepsilon > 0$ *error* **then**
            $k \leftarrow k + 1$, break;
        end
    end
end

Remark 7.2 (*Measurement uncertainties*) If the measurements x and $\dot{x}$ are subject to uncertainties and the uncertainties are bounded, then we can apply the bounds of x and $\dot{x}$ in evaluating t_{k+1} by (7.18) instead of x and $\dot{x}$ themselves. In other words, $e(t)$ and $\dot{e}(t)$ are determined by the bounds of x, $\dot{x}$ and the state values of the adaptive system (7.1).

7.2.2   High-Relative-Degree Constraints

In this subsection, we consider safety constraints (2.18) whose relative degree is larger than one with respect to the real unknown dynamics and (7.1). In other words, unlike the last subsection, we now need to consider the HOCBF constraint (3.8) to determine the state feedback control that can ensure the satisfaction of (2.18).

Similar to the last subsection, we define the error state e as in (7.2) and rewrite the HOCBF $b(x)$ as in (7.3). The HOCBF constraint (3.8) that guarantees $b(\bar{x} + e) \geq 0$ with respect to the real unknown dynamics is

$$\frac{\partial^m b(x)}{\partial \bar{x}^m} f_a^{[m]}(\bar{x}) + \frac{\partial^m b(x)}{\partial \bar{x}^m} f_a^{[m-1]}(\bar{x}) g_a^{[1]}(\bar{x}) u + \frac{\partial^m b(x)}{\partial e^m} e^{(m)} + O(b(x))$$
$$+ \alpha_m(\psi_{m-1}(x)) \geq 0, \tag{7.21}$$

where $\frac{\partial^m b(x)}{\partial \bar{x}^m} f_a^{[m]}(\bar{x})$ denotes the m-times partial derivative of $b(x)$ with respect to $\bar{x}$ along $f_a(\bar{x})$ (a similar concept to the Lie derivative in [26]), and we have similar definitions for $\frac{\partial^m b(x)}{\partial \bar{x}^m} f_a^{[m-1]} g_a^{[1]}(\bar{x})$. The term $O(b(x))$ also contains the remaining time derivatives of e with degree less than m, while $e^{(i)} = x^{(i)} - \bar{x}^{(i)}, i \in \{1, \ldots, m\}$ is the ith derivative evaluated on line through $x^{(i)}$ (from a sensor) of the real system and $\bar{x}^{(i)}$ is the one coming from (7.1).

In order to find a condition that guarantees the satisfaction of the last equation in $[t_i, t_{i+1})$, $i = 1, 2, \ldots$, we let e and $e^{(i)}, i \in \{1, \ldots, m\}$ be bounded by $w \in \mathbb{R}^n_{>0}$ and $v_i \in \mathbb{R}^n_{>0}$, i.e., we have

$$|e| \leq w, \qquad |e^{(i)}| \leq v_i, \quad i \in \{1, \ldots, m\}, \tag{7.22}$$

where the inequalities are interpreted componentwise and the absolute value function $|\cdot|$ applies to each component. We also consider the set of states defined in (7.11) and we further define a set $S_h(t_k)$:

$$S_h(t_k) = \{y, e, e^{(1)}, \ldots, e^{(m)} : y \in S(t_k), |e| \leq w,$$
$$\wedge_{i=1}^m (|e^{(i)}| \leq v_i), y + e \in \cap_{i=1}^m C_i\} \tag{7.23}$$

where $\wedge_{i=1}^m$ denotes the conjunction from 1 to m.

Then, we can find the minimum values $b_{f_a^m}^{min}(t_k) \in \mathbb{R}$, $b_{\alpha_m}^{min}(t_k) \in \mathbb{R}$, $b_{e^m}^{min}(t_k) \in \mathbb{R}$, $b_R^{min}(t_k) \in \mathbb{R}$ for the preceding time interval that satisfy $(y, e, e^{(1)}, \ldots, e^{(m)}) \in S_h(t_k)$ starting at time t_k by

$$b_{f_a^m}^{min}(t_k) = \min_{(y,e,e^{(1)} \ldots, e^{(m)}) \in S_h(t_k)} \frac{\partial^m b(y+e)}{\partial y^m} f_a^{[m]}(y) \tag{7.24}$$

$$b_{\alpha_m}^{min}(t_k) = \min_{(y,e,e^{(1)}, \ldots, e^{(m)}) \in S_h(t_k)} \alpha_m(\psi_{m-1}(y+e)) \tag{7.25}$$

$$b_{em}^{min}(t_k) = \min_{(y,e,e^{(1)},...,e^{(m)}) \in S_h(t_k)} \frac{\partial^m b(y+e)}{\partial e^m} e^{(m)} \tag{7.26}$$

$$b_R^{min}(t_k) = \min_{(y,e,e^{(1)},...,e^{(m)}) \in S_h(t_k)} O(b(y+e)) \tag{7.27}$$

If $\frac{\partial^m b(x)}{\partial \bar{x}^m} f_a^{[m-1]} g_a^{[1]}(\bar{x})$ is independent of $\bar{x}$ and e, then we do not need to find its limit value within the set $S_h(t_k)$; otherwise, we can determine $b_{g_i}^{lim}(t_k) \in \mathbb{R}, i \in \{1, \ldots, q\}$, under the assumption that $\frac{\partial^m b(x)}{\partial \bar{x}^m} f_a^{[m-1]} g_i^{[1]}(\bar{x})$ has the same sign for all $x \in X, \bar{x} \in X$ (the sign of $u_i(t_k), i \in \{1, \ldots, q\}, k = 1, 2 \ldots$ can also be determined by solving each CBF-based QP associated with (7.7) at t_k with (7.21)), by

$$b_{g_i}^{lim}(t_k) = \begin{cases} \min\limits_{(y,e,e^{(1)}...e^{(m)}) \in S_h(t_k)} \frac{\partial^m b(y+e)}{\partial y^m} f_a^{[m-1]} g_i^{[1]}(y), & \text{if } u_i(t_k) \geq 0, \\ \max\limits_{(y,e,e^{(1)}...e^{(m)}) \in S_h(t_k)} \frac{\partial^m b(y+e)}{\partial y^m} f_a^{[m-1]} g_i^{[1]}(y), & \text{otherwise} \end{cases} \tag{7.28}$$

Let $b_{g_a}^{lim}(t_k) = (b_{g_1}^{lim}(t_k), \ldots, b_{g_q}^{lim}(t_k)) \in \mathbb{R}^{1 \times q}$, and we set $b_{g_a}^{lim}(t_k) = \frac{\partial^m b(x)}{\partial \bar{x}^m} f_a^{[m-1]} g_a^{[1]}(\bar{x})$ if it is independent of $\bar{x}$ and e for notational simplicity.

Similar to Remark 7.1, we can break the above terms into smaller components and determine their corresponding minimum values in order to make this approach less conservative. The condition that guarantees the satisfaction of (7.21) in the time interval $[t_k, t_{k+1})$ is then given by

$$b_{f_a^m}^{min}(t_k) + b_{g_a}^{lim}(t_k) u(t_k) + b_{em}^{min}(t_k) + b_{\alpha_m}^{min}(t_k) + b_R^{min}(t_k) \geq 0. \tag{7.29}$$

In order to apply the above condition to problem (7.7), we just replace (7.6) by (7.29), i.e., we have

$$\min_{u(t),\delta(t)} \int_0^T [||u(t)||^2 + p\delta^2(t)] dt \tag{7.30}$$

subject to (7.29), (2.19) and (7.8).

Based on the above, we once again define three events that trigger a solution of each QP associated with (7.30):

- **Event 1**: $|e| \leq w$ is about to be violated.
- **Event 2**: $|e^{(i)}| \leq v_i$ is about to be violated for each $i \in \{1, \ldots, m\}$.
- **Event 3**: the state of (7.1) reaches the boundaries of $S(t_k)$.

The next triggering time instant $t_{k+1}, k = 1, 2 \ldots$ ($t_1 = 0$) to solve a QP is determined by:

$$\begin{aligned} t_{k+1} = \min\{t > t_k : |e(t)| = w \text{ or } |e^{(i)}(t)| = v_i, \\ i \in \{1, \ldots, m\} \text{ or } |\bar{x}(t) - \bar{x}(t_k)| = s\} \end{aligned} \tag{7.31}$$

Formally, we have the following theorem that shows the satisfaction of the safety constraint (2.18) for the real unknown dynamics:

> **Theorem 7.2** ([75]) *Given a HOCBF $b(x)$ as in Definition 3.3. Let $t_{k+1}, k = 1, 2 \ldots$ be determined by (7.31) with $t_1 = 0$, and $b_{f_m^{a}}^{min}(t_k)$, $b_{\alpha_m}^{min}(t_k)$, $b_{e_m}^{min}(t_k)$, $b_R^{min}(t_k)$, $b_{g_a}^{lim}(t_k)$ be determined by (7.24)–(7.28), respectively. Then, under Assumption 7.1, any control $u(t_k)$ that satisfies (7.29) and updates the real unknown dynamics and (7.1) within time interval $[t_k, t_{k+1})$ renders the set $C_1 \cap \cdots \cap C_m$ forward invariant for the real unknown dynamics.*

Proof Similar to the proof of Theorem 7.1, we have

$$\frac{\partial^m b(x(t))}{\partial t^m} + O(b(x(t))) + \alpha_m(\psi_{m-1}(x(t))) \geq 0, \forall t \in [t_k, t_{k+1}], k = 1, 2 \ldots$$

which is equivalent to the HOCBF constraint (3.8) in Definition 3.3. Then, by Theorem 3.2, (7.3) and $e^{(i)} = x^{(i)} - \bar{x}^{(i)}$, $i \in \{1, \ldots, m\}$, we can recursively show that C_i, $i \in \{1, \ldots, m\}$ are forward invariant, i.e., the set $C_1 \cap \cdots \cap C_m$ is forward invariant for the real unknown dynamics. $\qquad\square$

This event-driven control process can be summarized through an algorithm similar to Algorithm 1. Measurement uncertainties can be dealt with as in Remark 7.2. We can also synchronize the state of the real unknown dynamics and (7.1) as in (7.19) and (7.20) such that we always have $e(t_k) = 0$ and $e^{(i)}(t_k)$, $i \in \{1, \ldots, m\}$ stays close to 0.

7.2.3 Extension to Multi-agent Systems

The event-driven control scheme presented thus far is based on a system with general dynamics as in (2.11). In the case of a multi-agent system [78], we consider a controlled agent (state $x \in X$ and control $u \in U$) whose dynamics are unknown, and a set S_a of other agents (state $y_i \in X$ for agent $i \in S_a$) whose dynamics are also unknown. For instance, the controlled agent could be the ego vehicle in autonomous driving, and other agents are either other vehicles or obstacles. The controlled agent is assumed to have on-board sensors to monitor its own state and that of other agents. For the unknown dynamics of the controlled agent and other agents, we adopt Assumption 7.1, i.e., we assume that the relative degree of each component of x is known with respect to the real unknown dynamics and the same applies to y_i, $i \in S_a$.

The analysis proceeds exactly as in Sects. 7.2.1 and 7.2.2 by defining $z := (y_1, e_x, \dot{e}_x, y_2, e_i, \dot{e}_i)$, where $y_1 \in S_x(t_k)$, $y_2 \in S_{y,i}(t_k)$, e_x pertains to the error state of the controlled agent, and e_i to the error state of the remaining agents $i \in S_a$. We also define an overall set $S(t_k)$:

$$S(t_k) = \{z \in \mathbb{R}^{6n} : y_1 \in S_x(t_k), |e_x| \leq w, |\dot{e}_x| \leq v, y_2 \in S_{y,i}(t_k), |e_i| \leq W_i,$$
$$|\dot{e}_i| \leq V_i, (y_1 + e_x, y_2 + e_i) \in C_1\}. \tag{7.32}$$

where the definitions of $S_x(t_k)$ and $S_{y,i}(t_k)$ are similar to the one in (7.11).

In this case, the number of events required to trigger a QP solution in the corresponding optimal control problem is increased from three in Sect. 7.2.1 to seven defined as follows:

- **Event 1**: $|e_x| \leq w$ is about to be violated.
- **Event 2**: $|\dot{e}_x| \leq v$ is about to be violated.
- **Event 3**: the state of the controlled agent reaches the boundaries of $S_x(t_k)$.
- **Event 4**: $|e_i| \leq W_i$ is about to be violated.
- **Event 5**: $|\dot{e}_i| \leq V_i$ is about to be violated.
- **Event 6**: the state of any agent $i \in S_a$ reaches the boundaries of $S_{y,i}(t_k)$.
- **Event 7**: $\frac{\partial b(x, y_i)}{\partial \bar{x}} g_j(\bar{x})$, $j \in \{1, \ldots, q\}$ changes sign for $t > t_k$ compared to the sign it had at t_k.

A theorem similar to Theorems 7.1 or 7.2 can then be established and the process in Algorithm 1 can be directly extended to the seven events above (and possibly an event that indicates a sign change in the u term in (7.6)). For more details, the reader is referred to [78].

7.3　ACC with Unknown Vehicle Dynamics

In this section, we revisit the version of the ACC problem considered in Example 4.3. All computations and simulations were conducted in MATLAB. We used quadprog to solve the quadratic programs and ode45 to integrate the dynamics.

In this case, the real vehicle dynamics are **unknown** to the controller:

$$\begin{bmatrix} \dot{v}(t) \\ \dot{z}(t) \end{bmatrix} = \begin{bmatrix} \sigma_1(t) + \frac{\sigma_3(t)}{M} u(t) - \frac{1}{M} F_r(v(t)) \\ \sigma_2(t) + v_p - v(t) \end{bmatrix} \tag{7.33}$$

where, as before, $x = (v, z)$ and $z(t)$ denotes the distance between the preceding and the ego vehicle, $v_p > 0$ and $v(t)$ denote the velocities of the preceding and ego vehicles along the same lane (the velocity of the preceding vehicle is assumed constant), respectively, and $u(t)$ is the control of the ego vehicle. In addition, $\sigma_1(t), \sigma_2(t), \sigma_3(t)$ denote three random processes whose pdf's have finite support. M denotes the mass of the ego vehicle and $F_r(v(t))$ denotes the resistance force, which is expressed [26] as: $F_r(v(t)) = f_0 sgn(v(t)) + f_1 v(t) + f_2 v^2(t)$, where $f_0 > 0$, $f_1 > 0$ and $f_2 > 0$ are unknown.

The adaptive dynamics will be automatically updated as shown in (7.20), and are in the form:

$$\underbrace{\begin{bmatrix} \dot{\bar{v}}(t) \\ \dot{\bar{z}}(t) \end{bmatrix}}_{\dot{\bar{x}}(t)} = \underbrace{\begin{bmatrix} h_1(t) - \frac{1}{M} F_n(\bar{v}(t)) \\ h_2(t) + v_p - \bar{v}(t) \end{bmatrix}}_{f_a(\bar{x}(t))} + \underbrace{\begin{bmatrix} \frac{1}{M} \\ 0 \end{bmatrix}}_{g_a(\bar{x}(t))} u(t) \tag{7.34}$$

where $h_1(t) \in \mathbb{R}$, $h_2(t) \in \mathbb{R}$ denote the two adaptive terms in (7.20), $h_1(0) = 0$, $h_2(0) = 0$. $\bar{z}(t)$, $\bar{v}(t)$ are corresponding to $z(t)$, $v(t)$ in (7.33). Note that $F_n(\bar{v}(t)) = g_0 sgn(\bar{v}(t)) + g_1 \bar{v}(t) + g_2 \bar{v}^2(t)$, which is *different* from F_r in (7.33), where $g_0 > 0$, $g_1 > 0$ and $g_2 > 0$ are empirically determined.

The control bound is defined as $-c_d Mg \leq u(t) \leq c_a Mg$, where $c_a > 0$ and $c_d > 0$ are the maximum acceleration and deceleration coefficients, respectively, and g is the gravity constant. We require that the distance $z(t)$ between the ego vehicle (real dynamics) and its immediately preceding vehicle be greater than $l_p > 0$, i.e.,

$$z(t) \geq l_p, \quad \forall t \geq 0. \tag{7.35}$$

The objective is to minimize $\int_0^T ((u(t) - F_r(v(t)))/M)^2 dt$. The ego vehicle is also trying to achieve a desired speed $v_d > 0$, which is implemented by a CLF $V(\bar{x}) = (\bar{v} - v_d)^2$ as in Definition 2.4. Since the relative degree of the constraint (7.35) is two, we define an HOCBF $b(x) = z - l_p$ with $\alpha_1(b(x)) = b(x)$ and $\alpha_2(\psi_1(x)) = \psi_1(x)$ as in Definition 3.3 to implement the safety constraint. Then, the HOCBF constraint (3.8) in this case is (with respect to the real dynamics (7.33)): $\ddot{b}(x) + 2\dot{b}(x) + b(x) \geq 0$. Combining (7.2), (7.34) and this equation, we have an HOCBF constraint in the form:

$$-h_1(t) + \frac{F_n(\bar{v}(t))}{M} + \frac{-1}{M}u(t) + \ddot{e}_2(t) + 2(h_2(t) + v_p$$
$$-\bar{v}(t) + \dot{e}_2(t)) + \bar{z}(t) + e_2(t) - l_p \geq 0 \tag{7.36}$$

where $e = (e_1, e_2)$, $e_1 = v - \bar{v}$, $e_2 = z - \bar{z}$.

Similar to (7.9), (7.10), we consider the state and bound the errors at step t_k, $k = 1, 2 \ldots$ for the above HOCBF constraint in the form: $\bar{v}(t_k) - s_1 \leq \bar{v} \leq \bar{v}(t_k) + s_1$, $\bar{z}(t_k) - s_2 \leq \bar{z} \leq \bar{z}(t_k) + s_2$, $|e_2| \leq w_2$, $|\dot{e}_2| \leq v_{2,1}$, $|\ddot{e}_2| \leq v_{2,2}$, where $s_1 > 0$, $s_2 > 0$, $w_2 > 0$, $v_{2,1} > 0$, $v_{2,2} > 0$.

As in (7.11), (7.20), we also synchronize the state and update the adaptive dynamics (7.34) at step t_k, $k = 1, 2 \ldots$ in the form:

$$\bar{v}(t_k) = v(t_k), \quad \bar{z}(t_k) = z(t_k),$$

$$h_1(t^+) = h_1(t^-) - \sum_{i=0}^{k} \ddot{e}_2(t_i), \quad h_2(t^+) = h_2(t^-) + \sum_{i=0}^{k} \dot{e}_2(t_i), \tag{7.37}$$

where

$$\dot{e}_2(t_k) = \dot{z}(t_k) - (h_2 + v_p - \bar{v}(t_k)),$$

$$\ddot{e}_2(t_k) = \ddot{z}(t_k) - \frac{F_n(\bar{v}(t_k)) - u(t_k^-)}{M} + h_1(t_k),$$

$u(t_k^-) = u(t_{k-1})$ and $u(t_0) = 0$. $\dot{z}(t_k)$, $\ddot{z}(t_k)$ are estimated by a sensor that measures the dynamics (7.33) at t_k.

Then, we can find the limit values using (7.12)–(7.15), solve a sequence of QPs associated with (7.7) at each time step t_k, $k = 1, 2 \ldots$, and evaluate the next time step t_{k+1} as in (7.18) afterwards. In the evaluation of t_{k+1}, we have $e_2 = z - \bar{z}$, $\dot{e}_2 = \dot{z} - (h_2 + v_p - \bar{v})$, $\ddot{e}_2 = \ddot{z} - \frac{F_n(\bar{v}) - u(t_k)}{M} + h_1$, where $z, \dot{z}, \ddot{z}$ are estimated by a sensor that measures the ego real dynamics (7.33), and $u(t_k)$ is already obtained by solving the QP and is held constant until we evaluate t_{k+1}. The optimizations similar to (7.12)–(7.15) are either QPs or LPs. Each QP or LP can be solved with a computational time $< 0.01s$ in MATLAB (Intel(R) Core(TM) i7-8700 CPU @ 3.2 GHz×2).

The simulation parameters are the same as those in Sect. 4.3. The pdf's of $\sigma_1(t)$, $\sigma_2(t)$, $\sigma_3(t)$ are uniform over the intervals $[-0.2, 0.2]$ m/s², $[-2, 2]$ m/s, $[0.9, 1]$, respectively. The sensor sampling rate is 20 Hz. We compare the proposed event-driven framework with the time-driven approach, where the discretization time for the time-driven approach is $\Delta t = 0.1$.

The simulation results are shown in Fig. 7.2a and b. In the event-driven approach (blue lines), the control varies largely in order to be responsive to the random processes in the real dynamics. If we decrease the uncertainty levels by a factor of 10, the control is smoother (magenta lines). Thus, highly accurately modeled adaptive dynamics are desired.

It follows from Fig. 7.2b that the set $C_1 \cap C_2$ is forward invariant for the real vehicle dynamics (7.33), i.e., the safety constraint (7.35) is guaranteed with the proposed event driven approach. In contrast, the safety is not guaranteed even with state synchronization under the time-driven approach.

In the event-driven approach, the number of QPs (events) within the interval $[0, T]$ is reduced by about 50% compared to the time-driven approach. If we multiply the bounds of the random processes $\sigma_1(t)$, $\sigma_2(t)$ by 2, then the number of events increases by about 23% for both 20 and 100 Hz sensor sampling rate, which shows that accurate adaptive dynamics can reduce the number of events, and thus improve computational efficiency.

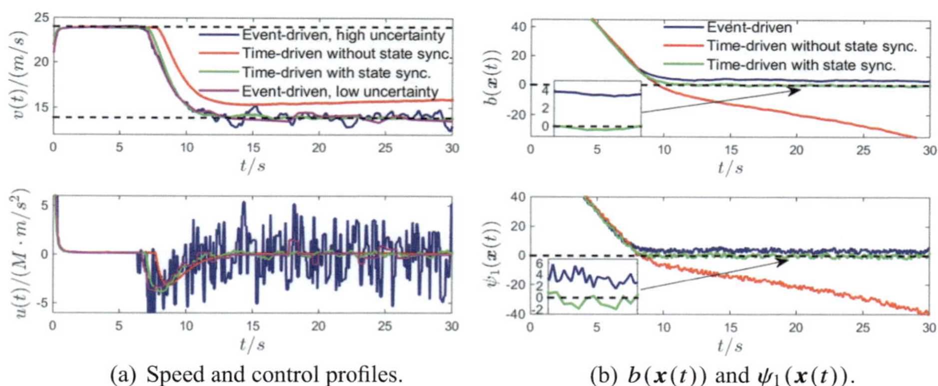

(a) Speed and control profiles.　　　　(b) $b(x(t))$ and $\psi_1(x(t))$.

Fig. 7.2 Results for the event-driven framework and the time-driven mechanism with or without synchronization. $b(x(t)) \geq 0$ and $\psi_1(x(t)) \geq 0$ imply the forward invariance of $C_1 \cap C_2$ (for the event-driven framework, but not for the time-driven case with or without synchronization)

Hierarchical Optimal Control with Barrier Functions

8

This chapter synthesizes controllers that *combine* optimal control and CBF methods, aiming for both optimality and guaranteed safety in real-time control. The motivation comes from the fact that analytical solutions for constrained optimal control problems are only possible for simple system dynamics and constraints. Moreover, the computational complexity for deriving such solutions significantly increases as one or more constraints become active and it grows as a power function of the number of constraints. Additional factors which further limit the real-time use of these methods include the presence of noise in the dynamics, model inaccuracies, environmental perturbations, and communication delays in the information exchange among system components. Thus, there is a gap between optimal control solutions (which represent a lower bound for the optimal achievable cost) and the execution of controllers aiming to achieve such solutions under realistic operational conditions.

The key idea in combining optimal control formulations with CBF methods is to first generate trajectories by solving a tractable optimal control problem and then seek to *track these trajectories* using a controller that simultaneously ensures that all state and control constraints are satisfied at all times. This is accomplished in two steps. The first step is to solve a constrained optimal control problem. Given a set of initial conditions, it is usually possible to derive simple conditions under which it can be shown that no constraint becomes active. In this case, executing the unconstrained optimal control solution becomes a relatively simple tracking problem. Otherwise, we can still often derive an optimal control solution consisting of both unconstrained and constrained arcs. However, such derivations may not always be feasible in real time. Either way, using the best possible analytical solution within reasonable real-time computational constraints (possibly just the unconstrained solution), this step leads to a reference control $u_{ref}(t)$, $t \in [t_0, t_f]$. The second step is then to use HOCBFs to account for constraints with arbitrary relative degrees, and define a sequence of QPs whose goal is to *optimally track* $u_{ref}(t)$ at each discrete time step over $[t_0, t_f]$. In

W. Xiao et al., *Safe Autonomy with Control Barrier Functions*,
Synthesis Lectures on Computer Science,
https://doi.org/10.1007/978-3-031-27576-0_8

this step, we can allow noise in the system dynamics and include nonlinearities which were ignored in the original optimal control solution. The resulting controller is termed *Optimal control with Control Barrier Functions (OCBF)*.

8.1   Problem Formulation and Approach

We begin with a class of constrained optimal control problems similar to that of Sect. 2.5, the main difference being that the terminal time of these problems is unspecified. This class of problems consists of an objective function, a set of safety requirements expressed as hard constraints on the system state, and a set of constraints on the control.

Objective: (*Cost minimization*) Consider an optimal control problem for a system with dynamics as in (2.11) and a cost defined as:

$$J = \int_{t_0}^{t_f} [\beta + C(x, u, t)]\, dt, \tag{8.1}$$

where t_0, t_f denote the initial and final times, respectively, and $C : \mathbb{R}^n \times \mathbb{R}^q \times [t_0, t_f] \to \mathbb{R}^+$ is a cost function. The parameter $\beta \geq 0$ is used to capture a trade-off between the minimization of the time interval $(t_f - t_0)$ and the operational cost $C(x, u, t)$. Keeping in mind that t_f is free (unspecified) and a decision variable in this problem, the terminal state t_f is constrained as follows.

Terminal state constraint: The state of system (2.11) is constrained to reach a point $\bar{X} \in X$, i.e.,

$$x(t_f) = \bar{X}, \tag{8.2}$$

Alternatively, this may also be considered as a soft constraint of the form $||x(t_f) - \bar{X}||^2$ added on to the cost in (8.1) as in (2.17).

Constraint 1 (*Safety constraints*): Let S_o denote an index set for a set of safety constraints. System (2.11) should always satisfy

$$b_j(x(t)) \geq 0, \quad \forall t \in [t_0, t_f], \tag{8.3}$$

where each $b_j : \mathbb{R}^n \to \mathbb{R}$, $j \in S_o$, is continuously differentiable.

Constraint 2 (*Control constraints*): These are provided by the control constraint set in (2.19).

Constraint 3 (*State constraints*): System (2.11) should always satisfy the state constraints (componentwise):

$$x_{\min} \leq x(t) \leq x_{\max}, \quad \forall t \in [t_0, t_f] \tag{8.4}$$

where $x_{\min} \in \mathbb{R}^n$ and $x_{\max} \in \mathbb{R}^n$. Note that we distinguish the state constraints above from the safety constraints in (8.3) since the latter are viewed as hard, while the former usually capture system capability limitations that can be relaxed to improve the problem feasibility; for example, in traffic networks, vehicles are constrained by upper and lower speed limits.

Thus, the problem we have formulated, labeled **OCP-F** (to capture the fact that the terminal time is "Free") can be summarized as follows:

OCP-F: Find a control policy for system (2.11) such that the cost (8.1) is minimized, and constraints (8.2), (8.3), (8.4) and (2.19) are strictly satisfied.

The cost in (8.1) can be properly normalized by defining

$$\beta := \frac{\alpha \, \sup_{x \in X, u \in U, \tau \in [t_0, t_f]} C(x, u, \tau)}{(1 - \alpha)} \tag{8.5}$$

where $\alpha \in [0, 1)$, and then multiplying (8.1) by $\frac{\alpha}{\beta}$. Thus, we construct a convex combination within the integrand as follows:

$$J = \int_{t_0}^{t_f} \left(\alpha + \frac{(1 - \alpha)C(x, u, t)}{\sup_{x \in X, u \in U, \tau \in [t_0, t_f]} C(x, u, \tau)} \right) dt. \tag{8.6}$$

If $\alpha = 1$, then we solve (8.1) as a minimum time problem. The normalized cost (8.6) facilitates a trade-off analysis between the two metrics, i.e., time and operational cost. However, we will use the simpler cost expression (8.1) throughout this paper. Thus, we can take $\beta \geq 0$ as a weight factor that can be adjusted to penalize time relative to the cost $C(x, u, t)$ in (8.1).

Our approach to solve **OCP-F** consists of two main steps:

Step 1: Analytically solve a simplified version of OCP-F. We use a standard Hamiltonian analysis to obtain an optimal control $u^*(t)$ and optimal state $x^*(t)$, $t \in [t_0, t_f]$ for the cost (8.1) and system (2.11), under the terminal state constraint (8.2), the safety constraints (8.3), and the control and state constraints (2.19), (8.4). However, in order to obtain a tractable analytical solution, we linearize or simplify the dynamics (2.11).

Step 2: Optimally track the solution from Step 1 subject to HOCBF constraints. There are usually unmodelled dynamics and measurement noise in (2.11) which were ignored in Step 1. Thus, we now consider a modified version of system (2.11) to denote the real dynamics:

$$\dot{x} = f(x) + g(x)u + w, \tag{8.7}$$

where $w \in \mathbb{R}^n$ denotes all unmodeled uncertainties in the dynamics. We consider x as a measured (observed) state which includes the effects of such unmodeled dynamics and measurement noise and which can be used in what follows. Allowing for the noisy dynamics (8.7), we set $u_{ref}(t) = u^*(t)$ (more generally, $u_{ref}(t) = h(u^*(t), x^*(t), x(t))$, $h : \mathbb{R}^q \times \mathbb{R}^n \times \mathbb{R}^n \to \mathbb{R}^q$) and use the HOCBF method to track the optimal control as a reference, i.e., we solve the problem

$$\min_{\boldsymbol{u}(t)} \int_{t_0}^{t_f} ||\boldsymbol{u}(t) - \boldsymbol{u}_{ref}(t)||^2 dt \tag{8.8}$$

subject to (i) the HOCBF constraints (3.8) corresponding to the safety constraints (8.3), (ii) the state constraints (8.4), and (iii) the control constraints (2.19).

Since we often treat the terminal state constraint (8.2) as a soft one and try to minimize the deviation $||\boldsymbol{x}(t_f) - \bar{\boldsymbol{X}}||^2$, we also define a CLF $V(\boldsymbol{x} - \boldsymbol{x}^*)$. Thus, the cost (8.8) is also subject to the corresponding CLF constraint (2.16). The resulting problem can then be solved by the approach described at the end of Chap. 2.

8.2   From Planning to Execution of Optimal Trajectories

In this section, we describe how to solve problem **OCP-F** by seeking optimality as defined at the planning stage (Step 1 described above) while providing safety guarantees which always apply at the execution stage (Step 2 described above).

8.2.1   Optimal Trajectory Planning

To carry out Step 1 of the OCBF approach, let us consider a properly linearized version of (8.7) without the noise $\boldsymbol{w}$:

$$\dot{\boldsymbol{x}} = A\boldsymbol{x} + B\boldsymbol{u}, \tag{8.9}$$

where $\boldsymbol{x} = (x_1, \ldots, x_n)$, $\boldsymbol{u} = (u_1, \ldots, u_q)$, $A \in \mathbb{R}^{n \times n}$, $B \in \mathbb{R}^{n \times q}$.

In order to construct the Hamiltonian for the minimization of (8.1) subject to constraints (8.2), (8.3), (8.4) and (2.19), let $\boldsymbol{\lambda}(t)$ be the costate vector corresponding to the state $\boldsymbol{x}$ in (8.9) and $\boldsymbol{b}(\boldsymbol{x})$ denote the vector obtained by concatenating all $b_j(\boldsymbol{x})$, $j \in S_o$. The Hamiltonian with the state constraints, control constraints and safety constraints adjoined (omitting time arguments for simplicity) is

$$H(\boldsymbol{x}, \boldsymbol{\lambda}, \boldsymbol{u}) = C(\boldsymbol{x}, \boldsymbol{u}, t) + \boldsymbol{\lambda}^T (A\boldsymbol{x} + B\boldsymbol{u}) + \boldsymbol{\mu}_a^T (\boldsymbol{u} - \boldsymbol{u}_{\max})$$
$$+ \boldsymbol{\mu}_b^T (\boldsymbol{u}_{\min} - \boldsymbol{u}) + \boldsymbol{\mu}_c^T (\boldsymbol{x} - \boldsymbol{x}_{\max}) + \boldsymbol{\mu}_d^T (\boldsymbol{x}_{\min} - \boldsymbol{x}) \tag{8.10}$$
$$- \boldsymbol{\mu}_e^T \boldsymbol{b}(\boldsymbol{x}) + \beta$$

The components of the Lagrange multiplier vectors $\boldsymbol{\mu}_a$, $\boldsymbol{\mu}_b$, $\boldsymbol{\mu}_c$, $\boldsymbol{\mu}_d$, $\boldsymbol{\mu}_e$ are positive when the constraints are active and become 0 when the constraints are strict.

Let us first assume that all the constraints (2.19), (8.3), (8.4) are not active in the time interval $[t_0, t_f]$. The Hamiltonian (8.10) then reduces to

$$H(\boldsymbol{x}, \boldsymbol{\lambda}, \boldsymbol{u}) = C(\boldsymbol{x}, \boldsymbol{u}, t) + \boldsymbol{\lambda}^T (A\boldsymbol{x} + B\boldsymbol{u}) + \beta \tag{8.11}$$

Observing that the terminal constraints (8.2) expressed as $\boldsymbol{\psi} := \boldsymbol{x} - \bar{\boldsymbol{X}} = 0$ are not explicit functions of time, the transversality condition [12] is

$$H(\boldsymbol{x}(t), \boldsymbol{\lambda}(t), \boldsymbol{u}(t))|_{t=t_f} = 0 \tag{8.12}$$

with $\boldsymbol{\lambda}(t_f) = [(\boldsymbol{v}^T \frac{\partial \boldsymbol{\psi}}{\partial \boldsymbol{x}})^T]_{t=t_f}$ as the costate boundary condition, where $\boldsymbol{v}$ denotes a vector of Lagrange multipliers. The Euler-Lagrange equations become:

$$\dot{\boldsymbol{\lambda}} = -\frac{\partial H}{\partial \boldsymbol{x}} = -\frac{\partial C(\boldsymbol{x}, \boldsymbol{u}, t)}{\partial \boldsymbol{x}} - A^T \boldsymbol{\lambda}, \tag{8.13}$$

and the necessary condition for optimality is

$$\frac{\partial H}{\partial \boldsymbol{u}} = \frac{\partial C(\boldsymbol{x}, \boldsymbol{u}, t)}{\partial \boldsymbol{u}} + B^T \boldsymbol{\lambda} = 0. \tag{8.14}$$

Given (8.11)–(8.14), the initial state of system (8.7), and the terminal constraint $\boldsymbol{x}(t_f) = \bar{\boldsymbol{X}}$, we can derive an *unconstrained* optimal state trajectory $\boldsymbol{x}^*(t)$ and optimal control $\boldsymbol{u}^*(t)$, $t \in [t_0, t_f]$, for problem **OCP-F**.

When one or more constraints in (2.19), (8.3), (8.4) become active in the time interval $[t_0, t_f]$, we use the interior point analysis [12] to determine the conditions that must hold on a constrained arc entry point and exit point (if one exists prior to t_f). We can then determine the optimal entry and exit points, as well as the constrained optimal control $\boldsymbol{u}^*(t)$ and optimal state trajectory $\boldsymbol{x}^*(t)$, $t \in [t_0, t_f]$. Depending on the computational complexity involved in deriving the complete constrained optimal solution, we can specify a planned reference control $\boldsymbol{u}_{ref}(t)$ and state trajectory $\boldsymbol{x}_{ref}(t)$, $t \in [t_0, t_f]$. For example, we may just plan for a safety-constrained solution and omit the state and control constraints (2.19), (8.4), or even plan for only the unconstrained optimal solution to simplify the trajectory planning process.

In summary, we obtain a solution $\boldsymbol{u}_{ref}(t)$ to a simplified version of problem **OCP-F** which is based on the linear system model (8.9) and as many constraints (possibly none) as our computational budget permits. As detailed next, we then proceed to optimally track $\boldsymbol{u}_{ref}(t)$ subject to appropriate HOCBF constraints which ensure the satisfaction of all original constraints.

8.2.2   Safety-Guaranteed Optimal Control with HOCBFs

In order to carry out Step 2 of the OCBF approach, we now introduce a method that tracks the planned optimal control and state trajectory while guaranteeing the satisfaction of all constraints (2.19), (8.3), (8.4) in problem **OCP-F**.

As explained in Sect. 8.2.1, we use $\boldsymbol{u}^*(t)$ and $\boldsymbol{x}^*(t)$, $t \in [t_0, t_f]$, to denote the optimal control and state trajectory derived under no active constraints or with some (or all) of the constraints active, depending on the associated computational complexity considered acceptable in a particular setting. We can then reformulate (8.1) as the following optimization problem in which t_f is fixed at the optimal terminal time determined along with $\boldsymbol{u}^*(t)$:

$$\min_{\boldsymbol{u}(t)} \int_{t_0}^{t_f} ||\boldsymbol{u}(t) - \boldsymbol{u}_{ref}(t)||^2 dt \tag{8.15}$$

subject to (2.19), (8.3), (8.4), where

$$\boldsymbol{u}_{ref}(t) = F_U(\boldsymbol{u}^*(t), \boldsymbol{x}^*(t), \boldsymbol{x}(t)) \tag{8.16}$$

is a specific function of the optimal control and state trajectory, as well as the actual state under noise $\boldsymbol{w}$ from (8.7). A typical choice for $F_U(\boldsymbol{u}^*(t), \boldsymbol{x}^*(t), \boldsymbol{x}(t))$ is

$$\boldsymbol{u}_{ref}(t) = e^{\sum_{j=1}^{n} \frac{x_j^*(t) - x_j(t)}{\sigma_j}} \boldsymbol{u}^*(t), \tag{8.17}$$

where $x_j(t)$, $j \in \{1, \dots, n\}$ denote the observed state variables under noise $\boldsymbol{w}$ from (8.7), $x_j^*(t)$, $j \in \{1, \dots, n\}$, $u_i^*(t)$, $i \in \{1, \dots, q\}$ denote the optimal state and control from the last subsection, and $\sigma_j > 0$, $j \in \{1, \dots, n\}$ are adjustable weight parameters. In (8.17), the sign of the term $x_j^*(t) - x_j(t)$ depends on whether $x_j(t)$ is increasing with $u_i(t)$. In particular, when $x_j(t) > x_j^*(t)$, for all $j \in \{1, \dots, n\}$, we have $u_i(t) < u_i^*(t)$ and the state errors can be automatically eliminated. If $x_j(t) < x_j^*(t)$, for all $j \in \{1, \dots, n\}$, the state errors can similarly be automatically eliminated. However, when $x_j(t) > x_j^*(t)$ and $x_{j+1}(t) < x_{j+1}^*(t)$, we may wish to enforce $u_i(t) < u_i^*(t)$, $i \in \{1, \dots, q\}$. Thus, it is desirable that $\sigma_j < \sigma_{j+1}$ (similarly, when $x_j(t) < x_j^*(t)$ and $x_{j+1}(t) > x_{j+1}^*(t)$). In summary, we select $\sigma_j > 0$, $j \in \{1, \dots, n\}$ such that $\sigma_j < \sigma_{j+1}$, $j \in \{1, \dots, n-1\}$.

Alternative forms of (8.16) include

$$\boldsymbol{u}_{ref}(t) = \sum_{j \in \{1, \dots, n\}} \frac{x_j^*(t)}{x_j(t)} \boldsymbol{u}^*(t) \tag{8.18}$$

and the state feedback tracking control approach from [26]:

$$\boldsymbol{u}_{ref}(t) = \boldsymbol{u}^*(t) + \sum_{j=1}^{n} k_j(x_j^*(t) - x_j(t)), \tag{8.19}$$

where $k_j > 0$, $\in \{1, \dots, n\}$. Clearly, there are several possible choices for the form of $\boldsymbol{u}_{ref}(t)$ which may depend on the specific application of interest.

We emphasize that the cost (8.15) is subject to all the constraints (2.19), (8.3), (8.4). We use HOCBFs to implement these constraints, as well as CLFs to better track the optimal state $\boldsymbol{x}^*(t)$, as shown in the following subsections.

8.2.2.1 Optimal State Tracking

First, we aim to track the optimal state $x^*(t)$ obtained in Sect. 8.2.1 using CLFs. We can always find a state variable x_k, $k \in \{1, \ldots, n\}$ in x that has relative degree one (assume x_k is the output) with respect to system (8.7). This is because we only take the Lie derivative of the Lyapunov function once in the CLF constraint (2.16). Then, we define a controller aiming to drive $x_k(t)$ to $x_{ref}(t)$ where $x_{ref}(t)$ is of the form

$$x_{ref}(t) = F_X(x^*(t), x(t)) \tag{8.20}$$

A typical choice analogous to (8.17) is

$$x_{ref}(t) = e^{\sum_{j \in \{1,\ldots,n\}\setminus k} \frac{x_j^*(t) - x_j(t)}{\sigma_j}} x_k^*(t) \tag{8.21}$$

where $\sigma_j > 0$, $j \in \{1, \ldots, n\} \setminus k$ and $\{1, \ldots, n\} \setminus k$ denotes excluding k from the set $\{1, \ldots, n\}$. An alternative form analogous to (8.18) is

$$x_{ref}(t) = \sum_{j \in \{1,\ldots,n\}\setminus k} \frac{x_j^*(t)}{x_j(t)} x_k^*(t) \tag{8.22}$$

where $x_j^*(t)$, $j \in \{1, \ldots, n\} \setminus k$ are the (unconstrained or constrained) optimal state trajectories from the Sect. 8.2.1, and $x_j(t) \neq 0$; otherwise, we can use (8.21). In (8.22), if $x_j(t) > x_j^*(t)$, then $x_{ref}(t) < x_k^*(t)$, thus automatically reducing (or eliminating) the tracking error. Note that while $x_{ref}(t)$ in (8.22) depends heavily on the exact value of $x_j(t)$, an advantage of (8.21) is that it allows $x_{ref}(t)$ to depend only on the error. Clearly, we can define different tracking forms instead of (8.22) and (8.21) depending on the specific characteristics of an application.

Using a specific selected form of $x_{ref}(t)$, we can now proceed as in Definition 2.4 and define an output $y_k(t) := x_k(t) - x_{ref}(t)$ for the state variable x_k which has relative degree one. Accordingly, we define a CLF $V(y_k(t)) = y_k^2(t)$ with $c_1 = c_2 = 1, c_3 = \epsilon > 0$ as in Definition 2.4. Then, any control input $u(t)$ should satisfy, for all $t \in [t_0, t_f]$,

$$L_f V(y_k(t)) + L_g V(y_k(t))u(t) + \epsilon V(y_k(t)) \leq \delta_k(t) \tag{8.23}$$

where $\delta_k(t)$ is a relaxation variable (to be minimized as explained in the sequel) enabling the treatment of the requirement $x_k(t) = x_{ref}(t)$ as a soft constraint. Note that we may also identify other state variables with relative degree one and define multiple CLFs to better track the optimal state. Note that (8.23) does not include any (unknown) noise term. Also note that selecting a larger ϵ can improve the state convergence rate [1].

8.2.2.2 Safety Constraints and State Limitations

Next, we use HOCBFs to map the safety constraints (8.3) and state limitations (8.4) from the state $x(t)$ to the control input $u(t)$. Let $b_j(x)$, $j \in S_o$, be the HOCBF corresponding to the jth safety constraint. In addition, let $b_{i,\max}(x) = x_{i,\max} - x_i$ and $b_{i,\min}(x) = x_i - x_{i,\min}$, $i \in \{1, \ldots, n\}$, be the HOCBFs for all state limitations, where $x_{\max} = (x_{1,\max}, \ldots, x_{n,\max})$, $x_{\min} = (x_{1,\min}, \ldots, x_{n,\min})$. The relative degrees of $b_{i,\max}(x)$, $b_{i,\min}(x)$, $i \in \{1, \ldots, n\}$ are m_i, and the relative degrees of $b_j(x)$, $j \in S_o$ are m_j. Therefore, in Definition 3.3, we choose HOCBFs with $m = m_i$ or m_j, including the penalty factors $p_{i,\min} > 0$, $p_{i,\max} > 0$, $p_{i,safe} > 0$ (see discussion after Definition 3.3) for all the class $\mathcal{K}$ functions. Following (3.8), any control input $u_i(t)$ should satisfy

$$L_f^{m_j} b_j(x) + L_g L_f^{m_j-1} b_j(x)u + O(b_j(x)) + p_{i,safe}\alpha_{m_j}(\psi_{m_j-1}(x)) \geq 0, \ j \in S_o, \quad (8.24)$$

$$L_f^{m_i} b_{i,\max}(x) + L_g L_f^{m_i-1} b_{i,\max}(x)u + O(b_{i,\max}(x)) + p_{i,\max}\alpha_{m_i}(\psi_{m_i-1}(x)) \geq 0, \quad (8.25)$$

$$L_f^{m_i} b_{i,\min}(x) + L_g L_f^{m_i-1} b_{i,\min}(x)u + O(b_{i,\min}(x)) + p_{i,\min}\alpha_{m_i}(\psi_{m_i-1}(x)) \geq 0, \quad (8.26)$$

for all $t \in [t_0, t_f]$, $i \in \{1, \ldots, n\}$. Note that $u \in U$ in (2.19) are already constraints on the control inputs, hence, we do not need to use HOCBFs for them.

8.2.2.3 Joint Optimal and HOCBF (OCBF) Controller

Using the HOCBFs and CLFs introduced in the last two subsections, we can reformulate problem (8.15) in the form:

$$\min_{u(t),\delta_k(t)} \int_{t_0}^{t_f} \left(\beta\delta_k^2(t) + ||u(t) - u_{ref}(t)||^2\right) dt, \quad (8.27)$$

subject to (8.7), (8.23), (8.24,), (8.25), (8.26), and (2.19), the initial conditions $x(t_0)$, and given t_0, t_f. Thus, we have combined the HOCBF method and the optimal control solution by using (8.16) to link the optimal state and control to $u_{ref}(t)$, and using (8.20) in the CLF $(x(t) - x_{ref}(t))^2$ to combine with (8.8). We refer to the resulting control $u(t)$ in (8.27) as the *OCBF control*.

Finally, in order to explicitly solve (8.27) as in Sect. 2.5.1, we partition the continuous time interval $[t_0, t_f]$ into equal time intervals $\{[t_0 + \omega\Delta t, t_0 + (\omega + 1)\Delta t)\}$, $\omega = 0, 1, 2, \ldots$. In each interval $[t_0 + \omega\Delta t, t_0 + (\omega + 1)\Delta t)$, we assume the control is constant and find a solution to the optimization problem in (8.27) using the CLF $y_k = (x_k(t) - x_{ref}(t))^2$ and associated relaxation variable $\delta_k(t)$. Specifically, at $t = t_0 + \omega\Delta t$ ($\omega = 0, 1, 2, \ldots$), we solve

$$\mathbf{QP} : \underset{t=t_0+\omega\Delta t}{} \ (u^\star(t), \delta_k^\star(t)) = \underset{u(t),\delta_k(t)}{\arg\min} \left[\beta\delta_k^2(t) + ||u(t) - u_{ref}(t)||^2\right] \quad (8.28)$$

subject to

$$A_{\text{clf}}[u(t), \delta_k(t)]^T \leq b_{\text{clf}} \quad (8.29)$$

$$A_{\text{cbf\_lim}}[\boldsymbol{u}(t), \delta_k(t)]^T \le b_{\text{cbf\_lim}} \tag{8.30}$$

$$A_{\text{cbf\_safe}}[\boldsymbol{u}(t), \delta_k(t)]^T \le b_{\text{cbf\_safe}} \tag{8.31}$$

The constraint parameters A_{clf}, b_{clf} pertain to the reference state tracking CLF constraint (8.23):

$$\begin{aligned} A_{\text{clf}} &= [L_g V(y_k(t)), \quad -1], \\ b_{\text{clf}} &= -L_f V(y_k(t)) - \epsilon V(y_k(t)). \end{aligned} \tag{8.32}$$

On the other hand, the constraint parameters $A_{\text{cbf\_lim}}$, $b_{\text{cbf\_lim}}$ capture the state HOCBF constraints (8.25) and the control bounds (2.19):

$$A_{\text{cbf\_lim}} = \begin{bmatrix} -L_g L_f^{m_i-1} b_{i,\max}(\boldsymbol{x}(t)), \ 0 \\ -L_g L_f^{m_i-1} b_{i,\min}(\boldsymbol{x}(t)), \ 0 \\ 1, \qquad\qquad 0 \\ -1, \qquad\qquad 0 \end{bmatrix},$$

$$b_{\text{cbf\_lim}} = \begin{bmatrix} L_f^{m_i} b_{i,\max}(\boldsymbol{x}) + O(b_{i,\max}(\boldsymbol{x})) + p_{i,\max}\alpha_{m_i}(\psi_{m_i-1}(\boldsymbol{x})) \\ L_f^{m_i} b_{i,\min}(\boldsymbol{x}) + O(b_{i,\min}(\boldsymbol{x})) + p_{i,\min}\alpha_{m_i}(\psi_{m_i-1}(\boldsymbol{x})) \\ \boldsymbol{u}_{max} \\ -\boldsymbol{u}_{min} \end{bmatrix}. \tag{8.33}$$

for all $i \in \{1, \ldots, n\}$. Finally, the constraint parameters $A_{\text{cbf\_safe}}$, $b_{\text{cbf\_safe}}$ capture the safety HOCBF constraints (8.24), for all $j \in S_o$:

$$\begin{aligned} A_{\text{cbf\_safe}} &= \left[-L_g L_f^{m_j-1} b_j(\boldsymbol{x}), \ 0 \right], \\ b_{\text{cbf\_safe}} &= L_f^{m_j} b_j(\boldsymbol{x}) + O(b_j(\boldsymbol{x})) + p_{j,safe}\alpha_{m_j}(\psi_{m_j-1}(\boldsymbol{x})). \end{aligned} \tag{8.34}$$

From a computational complexity point of view, it normally takes a fraction of a second to solve (8.28) in MATLAB, rendering the OCBF controller very efficient for real-time implementation. After solving each QP (8.28) we obtain an optimal OCBF control $\boldsymbol{u}^\star(t)$, not to be confused with a solution of the original optimal control problem (8.1). We then update (8.7) and apply it to all $t \in [t_0 + \omega\Delta t, t_0 + (\omega + 1)\Delta t]$. Several applications of the OCBF approach are given in Chap. 9.

Remark 8.1 If we can find conditions such that the constraints are not active [81], then we can simply track the unconstrained optimal control and state. This simplifies the implementation of the optimal trajectory planning without considering constraints, i.e., we can directly apply $\boldsymbol{u}_{ref}$ in (8.16) as the control input of system (8.7) instead of solving (8.28). The feasibility of QP (8.28) can be improved through smaller $p_{i,\min}$, $p_{i,\max}$, $p_{j,safe}$ at the expense of possibly shrinking the initial feasible set [71].

8.2.3 Constraint Violations Due to Noise

The presence of noise in the dynamics (8.7) will generally result in violations of the constraints (8.4) or (8.3), which prevents the HOCBF method from satisfying the forward invariance property [71]. Therefore, we must seek to minimize the time during which such a constraint is violated.

8.2.3.1 Relative Degree One Constraints

Suppose that a constraint $b(\boldsymbol{x}(t)) \geq 0$ (one of the constraints in (8.3) or (8.4)) has relative degree one for system (8.7). Let us first assume that $\boldsymbol{w}$ in (8.7) is bounded by $||\boldsymbol{w}|| \leq W$, where $W > 0$ is a scalar. Then, the following modified CBF constraint [30] can guarantee that $b(\boldsymbol{x}(t)) \geq 0$ is always satisfied under $||\boldsymbol{w}|| \leq W$:

$$L_f b(\boldsymbol{x}(t)) + L_g b(\boldsymbol{x}(t))\boldsymbol{u}(t) + \alpha(b(\boldsymbol{x}(t))) - \left\|\frac{db(\boldsymbol{x}(t))}{d\boldsymbol{x}}\right\| W \geq 0. \tag{8.35}$$

We may also consider

$$L_f b(\boldsymbol{x}(t)) + L_g b(\boldsymbol{x}(t))\boldsymbol{u}(t) + \alpha(b(\boldsymbol{x}(t))) - \left|\frac{db(\boldsymbol{x}(t))}{d\boldsymbol{x}}\right| W \geq 0. \tag{8.36}$$

if the noise is bounded in the form $||\boldsymbol{w}|| \leq W$, $W \geq \boldsymbol{0}$ (componentwise). The HOCBF constraint (3.8) with $m = 1$ is equivalent to $L_f b(\boldsymbol{x}(t)) + L_g b(\boldsymbol{x}(t))\boldsymbol{u}(t) + \alpha(b(\boldsymbol{x}(t))) + \frac{db(\boldsymbol{x}(t))}{d\boldsymbol{x}}\boldsymbol{w} \geq 0$ if we take the derivative of $b(\boldsymbol{x}(t))$ along the noisy dynamics (8.7). Thus, the satisfaction of (8.36) implies the satisfaction of this constraint. Note that the modified CBF constraint (8.36) is conservative since it always considers the (deterministic) noise bound $\boldsymbol{W}$.

Next, suppose a bound $\boldsymbol{W}$ is unknown, in which case we can proceed as follows. Assume the constraint is violated at time $t_1 \in [t_0, t_f]$ due to noise, i.e., we have $b(\boldsymbol{x}(t_1)) < 0$. We need to ensure that $b(\boldsymbol{x}(t))$ is strictly increasing after time t_1, i.e., $\dot{b}(\boldsymbol{x}(t)) \geq c(t)$, where $c(t)$ is positive and is desired to take the largest possible value maintaining the feasibility of the QP (8.28), i.e., we wish to maximize $c(t)$ at each time step (alternatively, we can set $c(t) = c > 0$ as a positive constant). Using Lie derivatives, we evaluate the change in $b(\boldsymbol{x}(t))$ along the flow defined by the state vector. Then, any control $\boldsymbol{u}(t)$ must satisfy

$$L_f b(\boldsymbol{x}(t)) + L_g b(\boldsymbol{x}(t))\boldsymbol{u}(t) \geq c(t) \tag{8.37}$$

since we wish to maximize $c(t)$ so that $b(\boldsymbol{x}(t))$ is strictly increasing even if the system is subject to the worst possible noise case. For this reason, in what follows we assume that the random process $\boldsymbol{w}(t)$ in (8.7) is characterized by a probability density function with finite support and we incorporate the maximization of $c(t)$ into the cost (8.27) as follows:

$$\min_{\boldsymbol{u}(t),\delta_k(t),c(t)} \int_{t_0}^{t_f} \left(\beta \delta_k^2(t) + ||\boldsymbol{u} - \boldsymbol{u}_{ref}||^2 - Kc(t)\right) dt, \tag{8.38}$$

where $K > 0$ is a large scalar weight parameter.

Note that several constraints may be violated at the same time. Starting from t_1, we apply the constraint (8.37) to the HOCBF optimizer instead of the HOCBF constraint (3.8), and $b(\boldsymbol{x}(t))$ will be positive again in finite time since it is strictly increasing. When $b(\boldsymbol{x}(t))$ becomes positive again at $t_2 \in [t_1, t_f]$, we can once again apply the HOCBF constraint (3.8).

8.2.3.2 High Relative Degree Constraints

If a constraint $b(\boldsymbol{x}(t)) \geq 0$ is such that $b : \mathbb{R}^n \to \mathbb{R}$ has relative degree $m > 1$ for (8.7), we can no longer find a modified CBF constraint as in (8.36) that guarantees $b(\boldsymbol{x}(t)) \geq 0$ under noise $\boldsymbol{w}$. This is because we need to know the bounds of the derivatives of $\boldsymbol{w}$ as $b(\boldsymbol{x}(t))$ will be differentiated m times. In other words, we need to recursively drive $b^{(i)}(\boldsymbol{x}(t)) = \frac{d^i b(\boldsymbol{x}(t))}{dt^i}$ to be positive from $i = m$ to $i = 1$ after it is violated at some time $t \in [t_0, t_f]$. Therefore, we need knowledge of the positive degree of $b(\boldsymbol{x}(t))$ at t which is defined as follows.

> **Definition 8.1** (*Positive degree*) The positive degree $\rho(t)$ of a relative degree m function $b : \mathbb{R}^n \to \mathbb{R}$ at time t is defined as:
>
> $$\rho(t) := \begin{cases} \min\limits_{i \in \{0,\ldots,m-1\}: b^{(i)}(\boldsymbol{x}(t)) > 0} i, & \text{if } \exists i \in \{0, \ldots, m-1\} \\ m & \text{otherwise} \end{cases} \tag{8.39}$$

If $b^{(i)}(\boldsymbol{x}(t)) \leq 0$, for all $i \in \{0, \ldots, m-1\}$, $\boldsymbol{u}(t)$ shows up in $b^{(m)}(\boldsymbol{x}(t))$ since the function b has relative degree m for system (8.7). Therefore, we may choose a proper control input $\boldsymbol{u}(t)$ such that $b^{(m)}(\boldsymbol{x}(t)) > 0$, and, in this case, $\rho(t) = m$. The positive degree of $b(\boldsymbol{x}(t))$ at time t is 0 if $b(\boldsymbol{x}(t)) > 0$.

Letting $\psi_0(\boldsymbol{x}, t) := b(\boldsymbol{x}(t))$, we can construct a sequence of functions $\psi_i : \mathbb{R}^n \to \mathbb{R}, \forall i \in \{1, \ldots, m\}$ similar to (3.3):

$$\psi_i(\boldsymbol{x}) := \begin{cases} \dot{\psi}_{i-1}, & \text{if } i < \rho(t), \\ \dot{\psi}_{i-1}(\boldsymbol{x}) - \varepsilon, & \text{if } i = \rho(t), \\ \dot{\psi}_{i-1}(\boldsymbol{x}) + \alpha_i(\psi_{i-1}(\boldsymbol{x})), & \text{otherwise.} \end{cases} \tag{8.40}$$

where $\alpha_i(\cdot), i \in \{1, \ldots, m\}$, denote class $\mathcal{K}$ functions of their argument and $\varepsilon > 0$ is a constant. We may choose $\varepsilon \geq \left| \frac{d\psi_{i-1}(\boldsymbol{x})}{d\boldsymbol{x}} \right| W$ if $\boldsymbol{w}$ is bounded as in (8.36).

We can then define a sequence of sets C_i similar to (3.4) associated with the $\psi_{i-1}(\boldsymbol{x}), i \in \{1, \ldots, m\}$ functions in (8.40). We replace the definitions of $\psi_{i-1}(\boldsymbol{x}), C_i, i \in \{1, \ldots, m\}$ in Definition 3.3 to define $b(\boldsymbol{x})$ to be a HOCBF.

If $\rho(t) = m$, then $\psi_m(\boldsymbol{x}(t)) = \dot{\psi}_{m-1}(\boldsymbol{x}(t)) - \varepsilon \geq 0$, which is equivalent to the HOCBF constraint (3.8). The control $\boldsymbol{u}$ that satisfies $\dot{\psi}_{m-1}(\boldsymbol{x}(t)) \geq \varepsilon > 0$ will drive $\psi_{m-1}(\boldsymbol{x}(t)) > 0$ in finite time. Otherwise, since $\psi_{\rho(t)}(\boldsymbol{x}(t)) > 0$ according to Definition 8.1, we can always choose proper class $\mathcal{K}$ functions $\alpha_i(\cdot), i \in \{\rho(t) + 1, \ldots, m\}$ such that $\psi_i(\boldsymbol{x}) \geq 0$, i.e., we can construct a non-empty set $C_{\rho(t)+1} \cap \cdots \cap C_m$ [71]. By Theorem 3.2, the set $C_{\rho(t)+1} \cap \cdots \cap C_m$ is forward invariant if the HOCBF constraint (3.8) is satisfied. In other words, $\psi_{\rho(t)}(\boldsymbol{x}(t)) \geq 0$ is guaranteed. Since $\psi_{\rho(t)}(\boldsymbol{x}(t)) = \dot{\psi}_{\rho(t)-1}(\boldsymbol{x}(t)) - \varepsilon$, then $\dot{\psi}_{\rho(t)-1}(\boldsymbol{x}(t)) \geq \varepsilon > 0$. The function $\psi_{\rho(t)-1}(\boldsymbol{x}(t))$ will become positive in finite time, and the positive degree of $b(\boldsymbol{x}(t))$ will decrease by one. Proceeding recursively at most m times, eventually the positive degree of $b(\boldsymbol{x}(t))$ will be 0, i.e., the original constraint $b(\boldsymbol{x}(t)) > 0$ is satisfied in finite time. The time needed for the constraint $b(\boldsymbol{x}(t)) > 0$ to be satisfied depends on the magnitude of ε.

In addition to the above approach, one may also consider finite time convergence CBFs [57, 76] or time-varying CBFs [31] to ensure the satisfaction of a constraint within specified time if it is initially violated.

Applications to Autonomous Vehicles in Traffic Networks

9

Advancements in next generation transportation system technologies have seen the emergence of Autonomous Vehicles (AVs) in traffic networks. When these AVs are further equipped with the ability to communicate with each other and exchange information, they are referred to as Connected Autonomous Vehicles or Connected and Automated Vehicles (CAVs). Their deployment promises dramatic improvements in terms of safety, reductions in congestion, energy consumption, and air pollution, as well as increased comfort for drivers and passengers [55, 64, 68].

The control and coordination of CAVs is particularly challenging at critical conflict areas of a transportation system such as merging points (usually, highway on-ramps), roundabouts and intersections. Clearly, *decentralized* control mechanisms are preferable, since all computation is performed on board each vehicle and shared only with a small number of other vehicles that are affected by it. Optimal control problem formulations are used in some of these approaches, while Model Predictive Control (MPC) techniques are employed as an alternative, primarily to account for additional constraints and to compensate for disturbances by re-evaluating optimal actions [41, 46].

An alternative to MPC is provided by the use of CBFs since they have the property of guaranteeing safety constraints which, in the case of CAVs, is the most crucial requirement for the acceptance and viability of long-awaited self-driving vehicles. In this chapter, we show how to apply the CBF-methods developed throughout the book to the coordination of CAVs in traffic networks, specifically exploiting the OCBF method presented in Chap. 8 [84, 86]. We consider three specific applications corresponding to traffic merging, roundabouts, and signal-free intersections.

   121
W. Xiao et al., *Safe Autonomy with Control Barrier Functions*,
Synthesis Lectures on Computer Science,
https://doi.org/10.1007/978-3-031-27576-0_9

9.1   Traffic Merging

In this section, we consider the problem of coordinating CAVs merging from roads, which are generally curved, with the goal of jointly minimizing their travel time and energy consumption as well as passenger discomfort, which results from centrifugal forces experienced at higher speeds during the merging process. The safety constraints include maintaining a speed-dependent safe distance for collision avoidance at the merging point and everywhere throughout the approaching area. The lateral rollover avoidance constraint is obtained through the Zero Moment Point (ZMP) [54] method that is usually used in balancing legged robots.

We should point out that the case of CAVs traveling and merging along straight roads was analyzed in [83] as an optimal control problem assuming simple linear dynamics for all CAVs. In this case, explicit analytical solutions can be derived, even when all constraints become active. The price to pay for such a complete analytical solution is the computational cost, which can range from under $0.1\,$s to several seconds whenever multiple constraints become active, in which case one has to derive optimal controls for all constrained arcs along an optimal trajectory. In the case of curved roads, such analytical solutions become intractable and we will proceed using the OCBF approach of Chap. 8 (see also [86]), where optimal control and CBFs are combined. In particular, we first derive an optimal solution when no constraints become active in the optimal control problem. Then, we employ the OCBF framework to optimally track this solution while also guaranteeing the satisfaction of all constraints. The OCBF framework also allows us to study the trade-off between travel time, centrifugal comfort and energy consumption. A simulation study of an actual curved road merging problem that arises in the Massachusetts Turnpike is included in Sect. 9.1.3 showing significant improvements in the performance of the OCBF controller compared to a baseline with human-driven vehicles.

9.1.1   Traffic Merging Control as an Optimal Control Problem

The merging problem arises when traffic must be joined from two different roads, usually associated with a main lane and a merging lane as shown in Fig. 9.1. We consider the case where all traffic consists of CAVs randomly arriving at the two curved roads joined at the Merging Point (MP) M where a lateral collision may occur. Each segment from the origin O or O' to the merging point M has a length L for both roads and radii $r_{main} > 0, r_{merg} > 0$ for the main and merging roads, respectively. These two segments over which CAVs may communicate with each other and exchange state information is called the *Control Zone* (CZ). CAVs do not overtake each other in the CZ, as each road consists of a single lane. A multi-lane merging problem has been studied in [85] (without road curvatures), in which case overtaking is included. Thus, the problem here can be extended to one with multiple

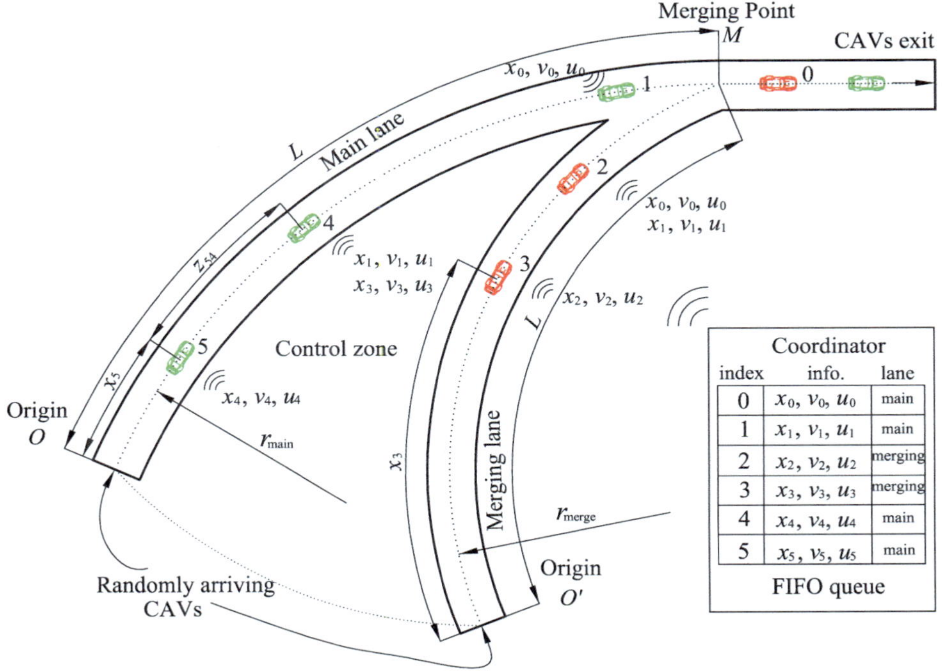

Coordinator		
index	info.	lane
0	x_0, v_0, u_0	main
1	x_1, v_1, u_1	main
2	x_2, v_2, u_2	merging
3	x_3, v_3, u_3	merging
4	x_4, v_4, u_4	main
5	x_5, v_5, u_5	main
FIFO queue		

Fig. 9.1 The merging problem for roads with curvature

lanes in each road along similar lines; in this case, there are multiple MPs, similar to the multi-lane intersection problem presented in Sect. 9.3.

A coordinator is associated with the MP whose function is to maintain a First-In-First-Out (FIFO) queue of CAVs based on their arrival time at the CZ and to enable real-time communication with the CAVs that are in the CZ, including the last one leaving the CZ. The FIFO assumption, imposed so that CAVs cross the MP in their order of arrival, is made for simplicity and often to ensure fairness, but can be relaxed through dynamic resequencing schemes (e.g., [82]). An explicit resequencing method is included in the sequel.

Let $S(t)$ be the set of FIFO-ordered (according to arrival times at O or O') indices of all CAVs located in the CZ at time t; this includes the CAV (whose index is 0 as shown in Fig. 9.1) that has just left the CZ. Let $N(t)$ be the cardinality of $S(t)$. Thus, if a CAV arrives at O or O' at time t, it is assigned the index $N(t)$. All CAV indices in $S(t)$ decrease by one when a CAV passes over the MP and the vehicle whose index is -1 is dropped.

The vehicle dynamics for each CAV $i \in S(t)$ along the lane to which it belongs take the form

$$\dot{x}_i(t) = v_i(t), \qquad \dot{v}_i(t) = u_i(t), \tag{9.1}$$

where $x_i(t)$ denotes the distance to the origin O (O') along the main (merging) lane if the vehicle i is located in the main (merging) lane, $v_i(t)$ denotes the velocity, and $u_i(t)$ denotes the control input (acceleration). We consider three objectives for each CAV subject to four constraints, as detailed next.

Objective 1 (*Minimizing travel time*): Let t_i^0 and t_i^m denote the time that CAV $i \in S(t)$ arrives at the origin O or O' and the merging point M, respectively. We wish to minimize the travel time $t_i^m - t_i^0$ for CAV i.

Objective 2 (*Minimizing energy consumption*): We also wish to minimize energy consumption for each $i \in S(t)$:

$$\min_{u_i(t)} \int_{t_i^0}^{t_i^m} C_i(u_i(t))dt, \tag{9.2}$$

where $C_i(\cdot)$ is a strictly increasing function of its argument, and it usually takes the quadratic form: $C_i(u_i(t)) = u_i^2(t)$. More detailed energy consumption models are possible (e.g., [24]) and are considered in [86].

Objective 3 (*Maximizing centrifugal comfort*): In order to minimize the centrifugal discomfort (or maximize the comfort), we wish to minimize the centrifugal acceleration

$$\min_{u_i(t)} \int_{t_i^0}^{t_i^m} \kappa(x_i(t))v_i^2(t)dt, \tag{9.3}$$

where $\kappa : \mathbb{R} \to \mathbb{R}^{\geq 0}$ is the curvature of the road at position x_i. The curvature $\kappa(x_i)$ has a sign which is determined by $\frac{1}{r(x_i)}$, where $r : \mathbb{R} \to \mathbb{R}$ is the radius of the road at x_i. Since we just wish to minimize the centrifugal acceleration, we ignore the sign and set $\kappa(x_i) \geq 0$ in what follows.

Constraint 1 (*Safety constraints*): Let i_p denote the index of the CAV which physically immediately precedes i in the CZ (if one is present). We require that the distance $z_{i,i_p}(t) :=$ $x_{i_p}(t) - x_i(t)$ be constrained by the speed $v_i(t)$ of CAV $i \in S(t)$ so that

$$z_{i,i_p}(t) \geq \varphi v_i(t) + \delta, \quad \forall t \in [t_i^0, t_i^m], \tag{9.4}$$

where φ denotes the reaction time (as a rule, $\varphi = 1.8$ is used, e.g., [67]). If we define z_{i,i_p} to be the distance from the center of CAV i to the center of CAV i_p, then δ is a constant determined by the length of these two CAVs (taken to be a constant over all CAVs for simplicity).

Constraint 2 (*Safe merging*): There should be enough safe space at the MP M for a CAV (which eventually becomes CAV 1, as shown in Fig. 9.1) to cut in, i.e.,

$$z_{1,0}(t_1^m) \geq \varphi v_1(t_1^m) + \delta. \tag{9.5}$$

Constraint 3 (*Vehicle limitations*): There are constraints on the speed and acceleration for each $i \in S(t)$:

$$v_{i,min} \leq v_i(t) \leq v_{i,max}, \quad \forall t \in [t_i^0, t_i^m],$$
$$u_{min} \leq u_i(t) \leq u_{max}, \quad \forall t \in [t_i^0, t_i^m], \tag{9.6}$$

where $v_{i,max} > 0$ and $v_{i,min} \geq 0$ denote the maximum and minimum speed allowed in the CZ, while $u_{min} < 0$ and $u_{max} > 0$ denote the minimum and maximum control, respectively.

Constraint 4 (*Lateral safety constraint*): Finally, there is a constraint on the centrifugal acceleration to avoid lateral rollover for each $i \in S(t)$:

$$\kappa(x_i(t))v_i^2(t) \leq \frac{w_i^h}{h_i} g, \quad \forall t \in [t_i^0, t_i^m], \tag{9.7}$$

where $w_i^h > 0$ denotes the half-width of the vehicle, $h_i > 0$ denotes the height of the center of gravity with respect to the ground, and g is the gravity constant. The above lateral safety constraint is obtained through the Zero Moment Point (ZMP) [54] method (assuming the road lateral slope is zero) that balances the CAV considering both gravity and inertia.

Problem Formulation. Our goal is to determine a control law to achieve objectives 1–3 subject to constraints 1–4 for each $i \in S(t)$ governed by the dynamics (9.1). We first choose $C_i(u_i(t)) = \frac{1}{2}u_i^2(t)$ in (9.2), noting that the OCBF method allows for more elaborate fuel consumption models, e.g., as in [24]; in what follows, we shall limit ourselves to this model. Normalizing each objective, and combining objectives 1, 2 and 3 with $\alpha_1 \in [0, 1], \alpha_2 \in [0, 1 - \alpha_1]$, we formulate the following optimal control problem for each CAV:

$$\min_{u_i(t),t_i^m} \int_{t_i^0}^{t_i^m} \left[\alpha_1 + \alpha_2 \frac{\kappa(x_i(t))v_i^2(t)}{\kappa_{max} v_{max}^2} + (1 - \alpha_1 - \alpha_2) \frac{\frac{1}{2}u_i^2(t)}{\frac{1}{2}u_{lim}^2} \right] dt, \tag{9.8}$$

subject to (9.1), (9.4), (9.5), (9.6), (9.7), the initial and terminal position conditions $x_i(t_i^0) = 0$, $x_i(t_i^m) = L$, and given t_i^0, v_i^0 (where v_i^0 denotes the initial speed), and $u_{lim} := \max\{u_{max}^2, u_{min}^2\}$. Observe that the merging time t_i^m is a decision variable in this problem. Finally, the weight factor $\alpha_1 \geq 0$, $\alpha_2 \geq 0$ can be adjusted to penalize travel time and comfort relative to the energy cost.

Multiplying (9.8) by $\frac{u_{lim}^2}{2(1-\alpha_1-\alpha_2)}$ and letting

$$\beta_1 = \frac{\alpha_1 u_{lim}^2}{2(1 - \alpha_1 - \alpha_2)}, \quad \beta_2 = \frac{\alpha_2 u_{lim}^2}{2(1 - \alpha_1 - \alpha_2)\kappa_{max} v_{max}^2} \tag{9.9}$$

we have a simplified and normalized version of (9.8):

$$\min_{u_i(t),t_i^m} \int_{t_i^0}^{t_i^m} \left[\beta_1 + \beta_2 \kappa(x_i(t)) v_i^2(t) + \frac{1}{2} u_i^2(t) \right] dt. \tag{9.10}$$

9.1.2 Decentralized Online Control

Note that (9.10) can be locally solved by each CAV i provided that there is some information sharing with only two other CAVs: CAV i_p which physically immediately precedes i and is needed in (9.4) and CAV $i - 1$ so that i can determine whether this CAV is located in the same road or not. With this information, CAV i can determine which of two possible cases applies: (i) $i_p = i - 1$, i.e., i_p is the CAV immediately preceding i in the FIFO queue (e.g., $i = 3$, $i_p = 2$ in Fig. 9.1), and (ii) $i_p < i - 1$, which implies that CAV $i - 1$ is in a different road from i (e.g., $i = 4$, $i_p = 1$, $i - 1 = 3$ in Fig. 9.1). It is now clear that we can solve problem (9.10) for any $i \in S(t)$ in a decentralized way in the sense that CAV i needs only its own local state information and state information from $i - 1$, as well as from i_p in case (ii) above. Observe that if $i_p = i - 1$, then (9.5) is a redundant constraint; otherwise, we need to separately consider (9.4) and (9.5). Therefore, we will analyze each of these two cases in what follows.

Assumption 1 The safety constraint (9.4), state constraints (9.6), and lateral safety constraint (9.7) are not active at t_i^0.

Since CAVs arrive randomly, this assumption cannot always be enforced. However, we can handle violations of Assumption 1 by foregoing optimality and simply controlling a CAV that violates it until all constraints become feasible within the CZ using the CBF method [86].

Under Assumption 1, we will start by analyzing the case of no active constraints. The analysis of the cases where one or more constraints become active is similar to the straight road merging problem studied in [83]. However, the computational time significantly increases with constrained optimal solutions, which prevents the optimal control from being implementable in real-world merging problems. Therefore, in what follows, we use CBFs to guarantee the satisfaction of all the constraints, and employ the OCBF framework from Chap. 8 to optimally track the tractable unconstrained optimal solutions.

9.1.2.1 Unconstrained Optimal Control

Let $\mathcal{X}_i(t) := (x_i(t), v_i(t))$ be the state vector of CAV i and $\lambda_i(t) := (\lambda_i^x(t), \lambda_i^v(t))$ be the costate vector (for simplicity, in the sequel we omit explicit time dependence when no ambiguity arises). The Hamiltonian associated with (9.10) with the state constraint, control constraint and safety constraint adjoined is

$$H_i(\mathcal{X}_i, \lambda_i, u_i) = \frac{1}{2}u_i^2 + \beta_2 \kappa(x_i)v_i^2 + \lambda_i^x v_i + \lambda_i^v u_i$$
$$+ \mu_i^a(u_i - u_{max}) + \mu_i^b(u_{min} - u_i)$$
$$+ \mu_i^c(v_i - v_{max}) + \mu_i^d(v_{min} - v_i) \tag{9.11}$$
$$+ \mu_i^e(x_i + \varphi v_i + \delta - x_{i_p})$$
$$+ \mu_i^f(\kappa(x_i)v_i^2 - \frac{w_i^h}{h_i}g) + \beta_1.$$

The Lagrange multipliers μ_i^a, μ_i^b, μ_i^c, μ_i^d, μ_i^e, μ_i^f are positive when the constraints are active and become 0 when the corresponding inequalities are strict. Note that when the safety constraint (9.4) becomes active, i.e., $\mu_i^e > 0$, the expression above involves $x_{i_p}(t)$. When $i = 1$, the optimal trajectory is obtained without this term, since (9.4) is inactive over all $[t_1^0, t_1^m]$. Thus, once the solution for $i = 1$ is obtained (based on the analysis that follows), x_1^* is a given function of time and available to $i = 2$. Based on this information, the optimal trajectory of $i = 2$ is obtained. Similarly, all subsequent optimal trajectories for $i > 2$ can be recursively obtained based on $x_{i_p}^*(t)$ with $i_p = i - 1$.

Since the terminal state constraint $\psi_{i,1} := x_i(t_i^m) - L = 0$ is not an explicit function of time, the transversality condition [12] is

$$H_i(\mathcal{X}_i(t), \lambda_i(t), u_i(t))|_{t=t_i^m} = 0, \tag{9.12}$$

with the costate boundary condition $\lambda_i(t_i^m) = [(v_{i,1}\frac{\partial \psi_{i,1}}{\partial \mathcal{X}_i})^T]_{t=t_i^m}$, where $v_{i,1}$ denotes a Lagrange multiplier.

The Euler-Lagrange equations become

$$\dot{\lambda}_i^x = -\frac{\partial H_i}{\partial x_i} = -\mu_i^e - \beta_2 \frac{\partial \kappa(x_i)}{\partial x_i}v_i^2 - \mu_i^f \frac{\partial \kappa(x_i)}{\partial x_i}v_i^2, \tag{9.13}$$

and

$$\dot{\lambda}_i^v = -\frac{\partial H_i}{\partial v_i} \tag{9.14}$$
$$= -\lambda_i^x - \mu_i^c + \mu_i^d - \varphi\mu_i^e - 2\beta_2\kappa(x_i)v_i - 2\mu_i^f\kappa(x_i)v_i,$$

and the necessary condition for optimality is

$$\frac{\partial H_i}{\partial u_i} = u_i + \lambda_i^v + \mu_i^a - \mu_i^b = 0. \tag{9.15}$$

Since the curvature $\kappa(x_i)$ of the road usually depends on the specific road configuration and it prevents the derivation of an explicit solution, we replace $\kappa(x_i)$ by the average (or possibly maximum) curvature $\hat{\kappa} \geq 0$ of the road in the CZ. In the case of no active constraints throughout an optimal trajectory, we also have $\mu_i^a = \mu_i^b = \mu_i^c = \mu_i^d = \mu_i^e = \mu_i^f = 0$. Applying (9.15), the optimal control input is given by $u_i + \lambda_i^v = 0$ and the Euler-Lagrange equation (9.14) yields $\dot{\lambda}_i^v = -\lambda_i^x - 2\beta_2\hat{\kappa}v_i$. In the case of no active constraints throughout an optimal trajectory, (9.13) implies $\lambda_i^x(t) = a_i$, where a_i is an integration constant. Combining the last two equations, we have $\dot{u}_i = a + 2\beta_2\hat{\kappa}v_i$.

Combining dynamics (9.1) with the above, we have

$$\ddot{v}_i = a + 2\beta_2\hat{\kappa}v_i. \tag{9.16}$$

We can solve this differential equation and get the explicit solution for the speed as

$$v_i^*(t) = b_i e^{\sqrt{2\beta_2\hat{\kappa}}t} + c_i e^{-\sqrt{2\beta_2\hat{\kappa}}t} - \frac{a_i}{2\beta_2\hat{\kappa}}, \tag{9.17}$$

where b_i, c_i are integration constants.

Consequently, we obtain the following optimal solution for the unconstrained problem:

$$u_i^*(t) = \sqrt{2\beta_2\hat{\kappa}}(b_i e^{\sqrt{2\beta_2\hat{\kappa}}t} - c_i e^{-\sqrt{2\beta_2\hat{\kappa}}t}), \tag{9.18}$$

$$x_i^*(t) = \frac{1}{\sqrt{2\beta_2\hat{\kappa}}}(b_i e^{\sqrt{2\beta_2\hat{\kappa}}t} - c_i e^{-\sqrt{2\beta_2\hat{\kappa}}t}) - \frac{a_i}{2\beta_2\hat{\kappa}}t + d_i, \tag{9.19}$$

where d_i is also an integration constant. In addition, we have the initial conditions $x_i(t_i^0) = 0$, $v_i(t_i^0) = v_i^0$ and the terminal condition $x_i(t_i^m) = L$. The costate boundary conditions and (9.15) offer us $u_i(t_i^m) = -\lambda_i^v(t_i^m) = 0$ and $\lambda_i(t_i^m) = (a_i, 0)$, therefore, the transversality condition (9.12) gives us an additional relationship:

$$\beta_1 + \beta_2\hat{\kappa}v_i^2(t_i^m) + a_i v_i(t_i^m) = 0. \tag{9.20}$$

Then, for each $i \in S(t)$, we need to solve the following five nonlinear algebraic equations to get a_i, b_i, c_i, d_i and t_i^m:

$$
\begin{aligned}
&b_i e^{\sqrt{2\beta_2\hat{\kappa}}t_i^0} + c_i e^{-\sqrt{2\beta_2\hat{\kappa}}t_i^0} - \frac{a_i}{2\beta_2\hat{\kappa}} = v_i^0, \\[2mm]
&\frac{1}{\sqrt{2\beta_2\hat{\kappa}}}\left(b_i e^{\sqrt{2\beta_2\hat{\kappa}}t_i^0} - c_i e^{-\sqrt{2\beta_2\hat{\kappa}}t_i^0}\right) - \frac{a_i}{2\beta_2\hat{\kappa}}t_i^0 + d_i = 0, \\[2mm]
&\frac{1}{\sqrt{2\beta_2\hat{\kappa}}}\left(b_i e^{\sqrt{2\beta_2\hat{\kappa}}t_i^m} - c_i e^{-\sqrt{2\beta_2\hat{\kappa}}t_i^m}\right) - \frac{a_i}{2\beta_2\hat{\kappa}}t_i^m + d_i = L, \\[2mm]
&\sqrt{2\beta_2\hat{\kappa}}\left(b_i e^{\sqrt{2\beta_2\hat{\kappa}}t_i^m} - c_i e^{-\sqrt{2\beta_2\hat{\kappa}}t_i^m}\right) = 0, \\[2mm]
&\beta_1 + \beta_2\hat{\kappa}\left(b_i e^{\sqrt{2\beta_2\hat{\kappa}}t_i^m} + c_i e^{-\sqrt{2\beta_2\hat{\kappa}}t_i^m} - \frac{a_i}{2\beta_2\hat{\kappa}}\right)^2 \\[2mm]
&\quad + a_i\left(b_i e^{\sqrt{2\beta_2\hat{\kappa}}t_i^m} + c_i e^{-\sqrt{2\beta_2\hat{\kappa}}t_i^m} - \frac{a_i}{2\beta_2\hat{\kappa}}\right) = 0.
\end{aligned}
\tag{9.21}
$$

Note that when $\beta_2 \to 0$ (i.e., $\alpha_2 \to 0$), the optimal control (9.18) degenerates to the case without any comfort consideration. In other words, the optimal control (9.18) is in linear form as in [83]: $\lim_{\beta_2 \to 0} u_i(t) = f_a t + f_b$, where $f_a = 2\beta_2 \hat{\kappa}(b_i + c_i)$, $f_b = \sqrt{2\beta_2 \hat{\kappa}}(b_i - c_i)$ following a Taylor series expansion.

The equations in (9.21) are usually hard to solve as there are too many exponential terms (it usually takes about one second using *solve* in Matlab to solve them). This motivates us to use a computationally efficient solution approach as shown in the next subsection.

9.1.2.2 Explicit Solution for Integration Constants

In this section, we show how to determine an explicit solution for the integration constants of the optimal control (9.18) which significantly reduces the computational complexity. By Lemma 2 in [83], we have $t_i^m - t_i^0 = t_j^m - t_j^0$ if $v_i^0 = v_j^0$, $i \in S(t)$, $j \in S(t)$ if $\beta_2 = 0$. This is also true if $\beta_2 > 0$ as the total travel time clearly does not depend on the arrival time t_i^0 of a CAV $i \in S(t)$. Moreover, observe that v_i^0 shows up only in the first of the five equations in (9.21). Therefore, given β_1, β_2, we can get the solution for t_i^m for some fixed $v_i^0 \in [v_{min}, v_{max}]$ by solving (9.21) off line with $t_i^0 = 0$. We can then construct a look-up table over a finite number of v_i^0 values constrained by $v_i^0 \in [v_{min}, v_{max}]$ and use a simple linear interpolation for any possible v_i^0 value actually observed. However, since v_i^0 is continuous and this approach may induce non-negligible errors, we choose instead to use regression (e.g., with polynomial or Gaussian kernels) to obtain the solution of t_i^m for all $v_i^0 \in [v_{min}, v_{max}]$ in the form:

$$t_i^m = t_i^0 + R(v_i^0), \tag{9.22}$$

where $R : \mathbb{R} \to \mathbb{R}$ denotes the regression model. Thus, we can immediately obtain t_i^m on line from (9.22) for any CAV arriving at time t_i^0 with speed v_i^0.

Since the last equation of (9.21) is used to determine t_i^m, which is already evaluated as discussed above, it remains to use the first four equations to determine a_i, b_i, c_i, d_i, which are all in linear form. Therefore, we get

$$
\begin{bmatrix} a_i \\ b_i \\ c_i \\ d_i \end{bmatrix} =
\begin{bmatrix}
-\dfrac{1}{2\beta_2\hat{\kappa}} & e^{\sqrt{2\beta_2\hat{\kappa}}t_i^0} & e^{-\sqrt{2\beta_2\hat{\kappa}}t_i^0} & 0 \\[2mm]
-\dfrac{t_i^0}{2\beta_2\hat{\kappa}} & \dfrac{e^{\sqrt{2\beta_2\hat{\kappa}}t_i^0}}{\sqrt{2\beta_2\hat{\kappa}}} & -\dfrac{e^{-\sqrt{2\beta_2\hat{\kappa}}t_i^0}}{\sqrt{2\beta_2\hat{\kappa}}} & 1 \\[2mm]
-\dfrac{t_i^m}{2\beta_2\hat{\kappa}} & \dfrac{e^{\sqrt{2\beta_2\hat{\kappa}}t_i^m}}{\sqrt{2\beta_2\hat{\kappa}}} & -\dfrac{e^{-\sqrt{2\beta_2\hat{\kappa}}t_i^m}}{\sqrt{2\beta_2\hat{\kappa}}} & 1 \\[2mm]
0 & e^{\sqrt{2\beta_2\hat{\kappa}}t_i^m} & -e^{-\sqrt{2\beta_2\hat{\kappa}}t_i^m} & 0
\end{bmatrix}^{-1}
\begin{bmatrix} v_i^0 \\ 0 \\ L \\ 0 \end{bmatrix} . \tag{9.23}
$$

The above equation is then the explicit solution for the four integration constants of (9.18), and it is much more computationally efficient than solving (9.21). The above matrix is invertible if there exists a solution in (9.21).

9.1.2.3 Constrained Optimal Control

When one or more constraints in the merging problem becomes active, we can use the standard interior point analysis [12] for determining all constrained arcs in an optimal trajectory and their times when they start, similar to the straight-road merging case with no comfort constraints shown in [83]. This leads to a complete constrained optimal control solution. However, this solution can become complicated when two or more constraints become active in an optimal trajectory and very time-consuming to obtain, hence possibly prohibitive for real-time implementation. It is for this reason that we resort to the CBF method to guarantee the satisfaction of all constraints while sacrificing some performance if some constraints become active. This method can be implemented on line for the merging problems studied in [86].

9.1.2.4 Joint Optimal Control and Control Barrier Function (OCBF) Method

In this section, we briefly review the OCBF approach from Chap. 8 (see also [86]) as it applies to our merging problem. The OCBF controller aims to track the OC solution (9.18)–(9.19) while satisfying all constraints (9.4), (9.5), and (9.6). Each of the constraints in (9.4), (9.5), (9.6), and (9.7) can be enforced by a CBF as discussed in previous chapters. In particular, each of the continuously differentiable *state* constraints is mapped onto another constraint on the *control* input such that the satisfaction of this new constraint implies the satisfaction of the original constraint, as detailed in Chaps. 2 and 3. In addition, a Control Lyapunov Function (CLF), as defined in Sect. 2.4, can also be used to track (stabilize) the optimal speed trajectory (9.17)

Therefore, the OCBF controller solves the following problem:

$$\min_{u_i(t),e_i(t)} J_i(u_i(t),\, e_i(t)) = \int_{t_i^0}^{t_i^m} \left(\beta e_i^2(t) + \frac{1}{2}(u_i(t) - u_{ref}(t))^2 \right) dt, \tag{9.24}$$

subject to the CBF constraints that enforce (9.4), (9.6), (9.5), and (9.7), and to the CLF constraint for tracking. Here, $\beta > 0$ and $e_i(t)$ is a relaxation variable in the CLF constraint. The obvious selection for the control (acceleration) reference signal is $u_{ref}(t) = u_i^*(t)$ given by (9.18). However, we can improve the tracking process using $x_i^*(t)$ from (9.19) by selecting instead:

$$u_{ref}(t) = \frac{x_i^*(t)}{x_i(t)} u_i^*(t). \tag{9.25}$$

Alternative choices of $u_{ref}(t)$ are also possible as shown in [84, 86].

We refer to the resulting control $u_i(t)$ in (9.24) as the "OCBF control". The solution to (9.24) is obtained through the familiar by now method first presented at the end of Chap. 2: we discretize the time interval $[t_i^0, t_i^m]$ with time steps of length Δ and solve (9.24) over $[t_i^0 + k\Delta, t_i^0 + (k+1)\Delta], k = 0, 1, \ldots$, with $u_i(t), e_i(t)$ as decision variables held constant over each such interval. Consequently, each such problem is a QP since we have a quadratic cost and a number of linear constraints on the decision variables at the beginning of each interval. The solution of each such problem gives $u_i^*(t_i^0 + k\Delta)$, $k = 0, 1, \ldots$, allowing us to update (9.1) in the kth time interval. This process is repeated until CAV i leaves the CZ.

9.1.2.5 Dynamic Resequencing

The FIFO assumption imposed on the curved-road merging problem can potentially decrease CAV performance as the main and merging roads may have different curvatures. This non-symmetric structure generally benefits from non-FIFO CAV ordering as observed in [82]. In order to relax the FIFO assumption, the Optimal Dynamic Resequencing (ODR) approach in [82] includes a step before a new CAV arrives at one of the origins where the CAV obtains the constrained optimal solution and uses the joint objective function to determine the passing order at the MP. This approach is computationally expensive and, as already seen, the constrained optimal solutions are harder to be found in the curved-road merging problem.

In order to improve the computational efficiency of this process, we will relax the requirement for *optimal* resequencing by considering only the travel time under the unconstrained optimal control (9.18) as the objective used to determine the passing order. Since this resequencing may not be the optimal policy, we refer to it as *Dynamic Resequencing* (DR). Specifically, if the MP arrival time t_i^m under the unconstrained optimal control (9.18) of a new arrival CAV $i \in S(t)$ satisfies

$$t_j^m - t_i^m \geq \varphi + \frac{\delta}{v_j^*(t_i^m)}, \tag{9.26}$$

for some $j \in S(t)$ such that $i_p < j < i$ (i.e., j is located at a different road from i), then the newly arriving CAV i overtakes CAV j under $v_j^*(t_i^m)$ which is the unconstrained optimal speed of j from (9.17) at time t_i^m. If such j is found in (9.26), its safe merging constraint will change after resequencing. We conclude by noting that this analysis is still subject to the problem of possibly infeasible QPs which we have studied throughout the book, starting with Chap. 4.

9.1.3 Simulation Results for Traffic Merging Control

We have selected a merging configuration that occurs in the US interstate highway I-90 (known as the Massachusetts Turnpike) in the Boston area, as shown in Fig. 9.2. All CAVs start to communicate with a coordinator (at the merging point) in the resequencing/connection zone as shown in the figure. In order to compare the performance of the OCBF controller used in this setting to a baseline consisting entirely of human-driven vehicles, we have used the Vissim microscopic multi-model traffic flow simulation tool. The car-following model in Vissim is based on [69] and simulates human psycho-physiological driving behavior.

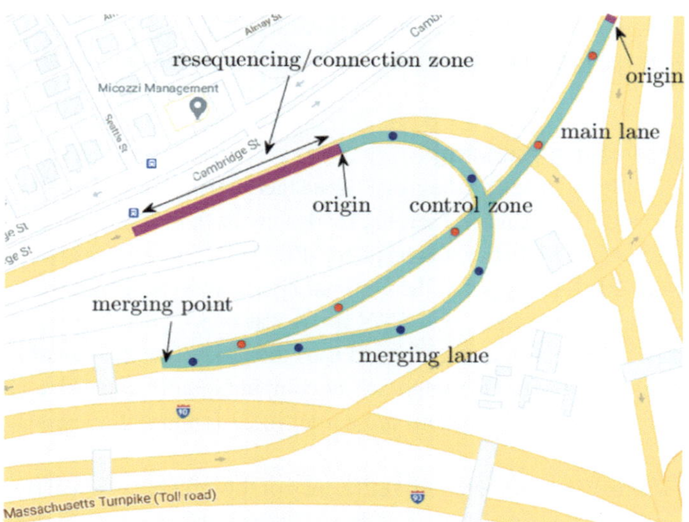

Fig. 9.2 A merging configuration at the I-90 Massachusetts Turnpike, USA

The parameters of the map shown in Fig. 9.2 are as follows: $L = 200$ m, $\hat{\kappa} = \frac{1}{200}$, $v_{i,max} = 20$ m/s in the main road, and $\hat{\kappa} = \frac{1}{50}$, $v_{i,max} = 15$ m/s in the merging road. In addition, $\Delta = 0.1$ s, $\varphi = 1.8$ s, $\delta = 0$ m, $v_{min} = 0$ m/s, $u_{max} = -u_{min} = 0.4$ g, $g = 9.81$ m/s^2. We set different cost weights for the main and merging roads, i.e., we set $\alpha_1 = 0.3$, $\alpha_2 = 0.1$ in the main road, and $\alpha_1 = 0.3$, $\alpha_2 = 0.4$ in the merging road. The simulation under optimal control is conducted in MATLAB by using the same arrival process input and initial conditions as in Vissim. The CAVs enter the CZ under a Poisson arrival process with an initial speed in the range $6.5-12.5$ m/s at the origins. The MATLAB computation time for the OCBF framework is minimal, i.e., less than 0.01 s for each QP of the OCBF controller (9.24) (Intel(R) Core(TM) i7-8700 CPU @ 3.2 GHz$\times 2$).

Simulation results comparing the OCBF controller to the VIssim car-following model under two different traffic rates are summarized in Table 9.1. Since we choose different α_1, α_2 for the main and merging roads, we list the metrics separately in Table 9.1.

When the traffic rates in the main and merging roads are equal at 500 CAVs/h, the overall objective function of CAVs in the main road improves about 22% with the OCBF method (using FIFO) compared with Vissim, and it further improves about 4% with the OCBF method when DR is included (see Sect. 9.1.2.5). Note that the energy consumption in the main road is 10 times worse for the OCBF method relative to Vissim, while the energy consumption in the merging road is 10 times better for the OCBF method. This is due to the fact that the CAVs in the main road have a higher speed limit than the ones in the merging road. Therefore, a CAV i in the main road may need a large energy consumption in order to satisfy the safe merging constraint at the MP when it has to coordinate with a CAV $i-1$

Table 9.1 Objective function comparison

Rate(CAVs/h)	Main:500, Merg.:500			Main:500, Merg.:800		
Method	Vissim	OCBF	DR	Vissim	OCBF	DR
Main time (s)	22.27	13.35	12.49	22.43	15.57	13.30
Main comfort	9.07	15.75	16.71	9.03	13.46	15.72
Main $\frac{1}{2}u_i^2(t)$	1.21	10.93	10.42	1.23	12.86	14.35
Main obj.	92.76	72.42	69.22	93.37	81.43	75.63
Merg. time (s)	26.71	15.92	15.94	39.02	16.40	16.32
Merg. comfort	35.89	50.77	50.73	30.61	49.50	49.75
Merg. $\frac{1}{2}u_i^2(t)$	13.78	0.05	1.13	13.29	1.79	3.20
Merg. obj.	301.3	238.4	239.6	606.6	241.0	242.3

whose speed is low in the merging road. Similar results are obtained when the arrival rate increases to 800 CAVs/h in the merging road.

When the traffic arrival rate increases to 1000 CAVs/h for both the main and merging roads, the human driven vehicles in the merging road will cause a heavy traffic congestion in Vissim, while the OCBF method can successfully manage the traffic without any congestion at all (see videos[1]).

9.2   Roundabout Traffic Control

Roundabouts are important components of a traffic network because they usually perform better than typical intersections in terms of efficiency and safety [20]. However, they can become significant bottleneck points as the traffic rate increases if traffic coordination is not properly executed, resulting in significant delays. Studies to date have mainly focused on conventional human-driven vehicles and have tried to solve the problem through improved road design or traffic signal control [38, 95, 101]. More recently, however, new methods have been proposed based on decentralized optimal control of CAVs in a roundabout.

In this section, we formulate an optimal control problem for controlling CAVs traveling through a roundabout as introduced in [97]. Unlike [14, 104], we *jointly* minimize the travel time and energy consumption and also consider speed-dependent safety constraints at a set of *Merging Points* (MPs) rather than merging zones (which makes solutions less conservative by improving roadway utilization). In addition, to improve computational efficiency, we adopt the joint Optimal Control and Control Barrier Function (OCBF) approach presented in Chap. 8 (see also [86]): we first derive an optimal solution when no constraints become active and subsequently optimally track this solution while also guaranteeing the satisfaction of all

---

[1] https://sites.google.com/view/xiaowei2021/research?authuser=0#h.o3pn0s7rv2ye.

constraints through the use of CBFs. Similar to the merging problem in Sect. 9.1, we initially assume a First-In-First-Out (FIFO) sequencing policy over the entire system. However, in this case the FIFO policy does not perform well in many "asymmetric" configurations. Thus, we also adopt an alternative sequencing policy, termed *Shortest Distance First* (SDF), which our experimental results show to be superior to FIFO. Nonetheless, the problem of systematically studying the effect of the passing order at roundabout MPs remains the subject of ongoing research.

9.2.1 Roundabout Traffic Control as an Optimal Control Problem

We consider a simplified single-lane triangle-shaped roundabout with 3 entries and 3 exits as shown in Fig. 9.3. The more common case of circular roundabouts has been recently analyzed in [98] based on similar ideas. We consider the case where all traffic consists of CAVs which randomly enter the roundabout from three different origins O_1, O_2 and O_3 and have assigned exit points E_1, E_2 and E_3. The gray road segments which include the triangle and three entry roads form the Control Zone (CZ) where CAVs can share information and thus be automatically controlled. We assume all CAVs move in a counterclockwise direction in the CZ. The entry road segments are connected with the triangle at the three Merging Points (MPs) where CAVs from different road segments may potentially collide with each other. The MPs are labeled M_1, M_2 and M_3. We assume that each road segment has one single lane (extensions to multiple lanes and MPs are possible following the analysis in [85]) The three entry road segments, which are labeled l_1, l_2 and l_3, have the same length L, while the road segments in the triangle, which are labeled l_4, l_5 and l_6, have the same length L_a (extensions to different lengths are straightforward). In Fig. 9.3, a circle, square and triangle represent entering from O_1, O_2 and O_3 respectively. The color red, green and blue represents exiting from E_1, E_2 and E_3 respectively. The full trajectory of a CAV in terms of the MPs it must go through can be determined by its entry and exit points.

A coordinator, i.e., a Road Side Unit (RSU) associated with the roundabout, maintains a passing sequence and records the information of each CAV. The CAVs communicate with the coordinator but are not controlled by it. All control inputs are evaluated on board each CAV in a *decentralized* way. Each CAV is assigned a unique index upon arrival at the CZ according to the passing order. The most common scheme for maintaining a passing sequence is the First-In-First-Out (FIFO) policy according to each CAV's arrival time at the CZ. The FIFO rule is one of the simplest schemes, yet works well in many situations as also shown in [93]. For simplicity, in what follows we use the FIFO policy, but we point out that any policy may be used, such as the Dynamic Resequencing (DR) method in [102]. We also introduce one such alternative in Sect. 9.2.2.

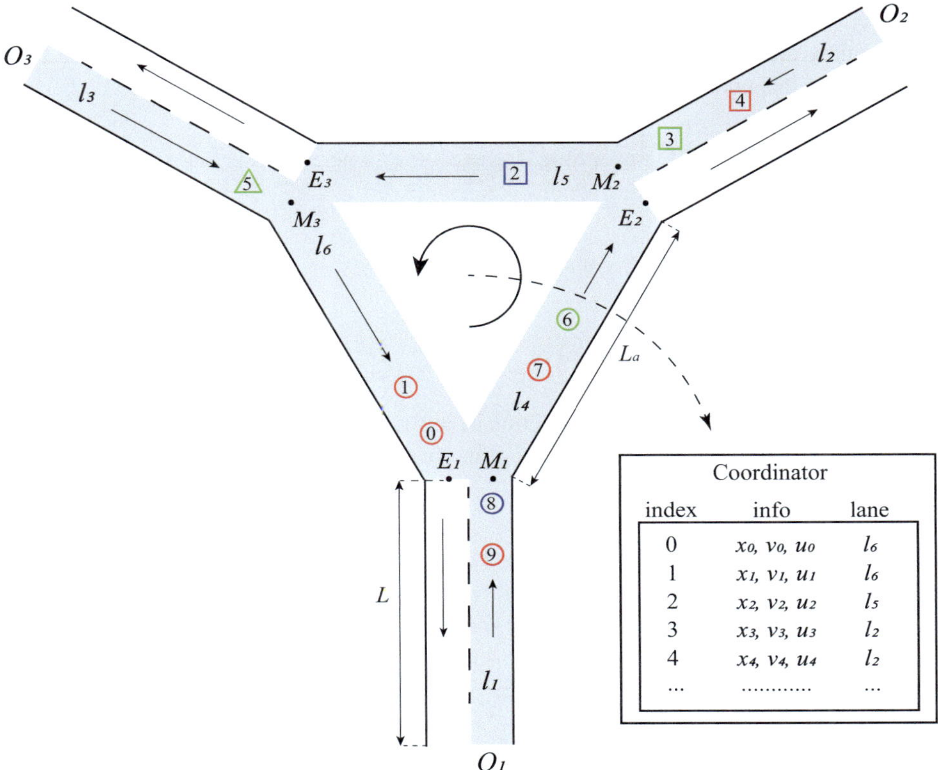

Fig. 9.3 A roundabout with 3 entries

Let $S(t)$ be the set of CAV indices in the coordinator queue table at time t. The cardinality of $S(t)$ is denoted as $N(t)$. When a new CAV arrives, it is allocated the index $N(t) + 1$. Each time a CAV i leaves the CZ, it is dropped and all CAV indices larger than i decrease by one. When CAV $i \in S(t)$ is traveling in the roundabout, there are several important *events* whose times are used in our analysis: (i) CAV i enters the CZ at time t_i^0, (ii) CAV i arrives at MP M_k at time t_i^k, $k \in \{1, 2, 3\}$, (iii) CAV i leaves the CZ at time t_i^f. Based on this setting, we can formulate an optimal control problem as described next.

Let $x_i^j(t)$ denote the distance from the origin O_j, $j \in \{1, 2, 3\}$ to the current location of CAV i along its trajectory as $x_i^j(t)$. Since the CAV's unique identity i contains the information about the CAV's origin O_j, we can use $x_i(t)$ instead of $x_i^j(t)$ (without any loss of information) to describe the vehicle dynamics as

$$\dot{x}_i(t) = v_i(t), \qquad \dot{v}_i(t) = u_i(t), \tag{9.27}$$

where v_i is the velocity CAV i along its trajectory and u_i is the acceleration (control input). We can now proceed similar to the traffic merging problem formulation in Sect. 9.1.1 to define an optimal control problem for each CAV as follows.

Objective 1 Minimize the travel time $J_{i,1} = t_i^f - t_i^0$ where t_i^0 and t_i^f are the times CAV i enters and exits the CZ.

Objective 2 Minimize energy consumption:

$$J_{i,2} = \int_{t_i^0}^{t_i^f} C_i(u_i(t))dt \tag{9.28}$$

where $C_i(\cdot)$ is a strictly increasing function of its argument. Since the energy consumption rate is a monotonic function of the acceleration, we adopt this general form to achieve energy efficiency.

Constraint 1 (*Rear-end safety constraint*) Let i_p denote the index of the CAV which immediately precedes CAV i on road segment l_k. The distance between i_p and i, $z_{i,i_p}(t) \equiv x_{i_p}(t) - x_i(t)$, should be constrained by a speed-dependent constraint:

$$z_{i,i_p}(t) \geq \varphi v_i(t) + \delta, \quad \forall t \in [t_i^0, t_i^f], \quad \forall i \in S(t) \tag{9.29}$$

where φ denotes the reaction time (as a rule, $\varphi = 1.8$ is suggested, see [67]), δ denotes the minimum safety distance (in general, we may use δ_i to make this distance CAV-dependent but will use a fixed δ for simplicity). The index of the preceding CAV's index i_p may change due to road segment changing events and is determined by the method described later in Sect. 9.2.2.2.

Constraint 2 (*Safe merging constraint*) Let t_i^k, $k \in \{1, 2, 3\}$ be the arrival time of CAV i at MP M_k. Let i_m denote the index of the CAV that CAV i may collide with when arriving at its next MP M_k. The distance between i_m and i, $z_{i,i_m}(t) \equiv x_{i_m}(t) - x_i(t)$, is constrained by:

$$z_{i,i_m}(t_i^k) \geq \varphi v_i(t_i^k) + \delta, \quad \forall i \in S(t), \quad k \in \{1, 2, 3\} \tag{9.30}$$

where i_m can be determined and updated by the method described in Sect. 9.2.2.2.

Constraint 3 (*Vehicle limitations*) The CAVs are also subject to velocity and acceleration constraints due to physical limitations or road rules:

$$v_{i,\min} \leq v_i(t) \leq v_{i,\max}, \forall t \in [t_i^0, t_i^f], \forall i \in S(t)$$
$$u_{i,\min} \leq u_i(t) \leq u_{i,\max}, \forall t \in [t_i^0, t_i^f], \forall i \in S(t) \tag{9.31}$$

where $v_{i,\max} > 0$ and $v_{i,\min} \geq 0$ denote the maximum and minimum speed for CAV i, $u_{i,\max} < 0$ and $u_{i,\min} < 0$ denote the maximum and minimum acceleration for CAV i. We further assume common speed limits dictated by the traffic rules at the roundabout, i.e. $v_{i,\min} = v_{\min}$, $v_{i,\max} = v_{\max}$.

As in the traffic merging problem, we construct a convex combination of the two objectives above:

$$J_i = \int_{t_i^f}^{t_i^0} \left[\alpha + (1-\alpha) \frac{\frac{1}{2} u_i^2(t)}{\frac{1}{2} \max\{u_{\max}^2, u_{\min}^2\}} \right] dt, \tag{9.32}$$

where $J_{i,1}$ and $J_{i,2}$ are combined with $\alpha \in [0, 1]$ after proper normalization. Here, we simply choose the quadratic function $C_i(u_i) = \frac{1}{2} u_i^2(t)$. If $\alpha = 1$, the problem degenerates into a minimum traveling time problem. If $\alpha = 0$, it degenerates into a minimum energy consumption problem.

By defining $\beta \equiv \frac{\alpha}{2(1-\alpha)} \max\{u_{\max}^2, u_{\min}^2\}$, $\alpha \in [0, 1)$ and proper scaling, we can rewrite this minimization problem as

$$J_i(u_i) = \beta(t_i^f - t_i^0) + \int_{t_i^0}^{t_i^f} \frac{1}{2} u_i^2(t) dt, \tag{9.33}$$

where β is the weight factor derived from α. This minimization is subject to constraints (9.27), (9.29), (9.30), (9.31), the initial condition $x_i(t_i^0) = 0$, and given t_i^0, v_i^0 and $x_i(t_i^f)$.

9.2.2   Decentralized Online Control

Compared to the single-lane merging control problem where the constraints are determined and fixed immediately when CAV i enters the CZ, the main difficulty in a roundabout is that the constraints generally change after every event (defined earlier). In particular, for each CAV i at time t only the merging constraint related to the next MP ahead is considered. In other words, we need to determine at most one i_p to enforce (9.29) and one i_m to enforce (9.30) at any time instant. There are two different ways to deal with this problem: (i) Treat the system as a single CZ with three MPs with knowledge of each CAV's sequence of MPs, or (ii) Decompose the roundabout into three separate merging problems corresponding to the three MPs, each with a separate CZ. The first approach is more complex and heavily relies on the CAV sequencing rule used. If FIFO is applied, it is likely to perform poorly in a large roundabout, because a new CAV may experience a large delay in order to preserve the FIFO passing sequence. The second approach is easier to implement but may cause congestion in a small roundabout due to the lack of space for effective control at each separate CZ associated with each MP.

Table 9.2 The extended coordinator queue table $S(t)$

$S(t)$

Idx	State	Curr.	Ori.	1st MP	2nd MP	3rd MP	i_p	i_m
0	x_0	l_6	l_1	M_1, M	M_2, M	M_3, M		
1	x_1	l_6	l_1	M_1, M	M_2, M	M_3, M	0	
2	x_2	l_5	l_2	M_2, M				
3	x_3	l_2	l_2	M_2	M_3	M_1		2
4	x_4	l_2	l_2	M_2	M_3		3	
5	x_5	l_3	l_3	M_3	M_1			1
6	x_6	l_4	l_1	M_1, M				
7	x_7	l_4	l_1	M_1, M	M_2	M_3	6	4
8	x_8	l_1	l_1	M_1	M_2			7
9	x_9	l_1	l_1	M_1	M_2	M_3	8	

In order to solve problem (9.33) for each CAV i, we need to first determine the corresponding i_p and i_m (when they exist) required in the safety constraints (9.29) and (9.30). Once this task is complete and (9.29) and (9.30) are fully specified, then this problem can be solved. In what follows, this first task is accomplished through a method designed to determine the constraints in an event-driven manner which can be used in either of the two approaches above and for any desired sequencing policy. An extended queue table, an example of which is shown in Table 9.2 corresponding to Fig. 9.3, is used to record the essential state information and identify all conflicting CAVs. We specify the state-updating mechanism for this queue table so as to determine for each CAV i the corresponding i_p and i_m. Then, we focus on the second approach above and develop a general algorithm for solving (9.33) based on the OCBF method.

9.2.2.1 The Extended Coordinator Queue Table

Starting with the coordinator queue table shown in Fig. 9.3, we extend it to include six additional columns for each CAV i. The precise definitions are given below:

- *idx*: Unique CAV index, which allows us to determine the order in which the CAV will leave the roundabout according to some policy (e.g., FIFO in Table 9.2 where rows are ordered by the index value).
- *state*: CAV state $x_i = (x_i, v_i)$ where x_i denotes the distance from the entry point to the location of CAV i along its current road segment.
- *curr.*: Current CAV road segment, which allows us to determine the rear-end safety constraints.

- *ori.*: Original CAV road segment, which allows us to determine its relative position when in road segment *curr*.
- *1st-3rd MP*: These columns record all the MPs on the CAV trajectory. If a CAV has already passed an MP, this MP is marked with an *M*. Otherwise, it is unmarked. As a CAV may not pass all three MPs in the roundabout, some of these columns may be blank.
- i_p: Index of the CAV that immediately precedes CAV i in the same road segment (if such a CAV exists).
- i_m: Index of the CAV that may conflict with CAV i at the next MP. CAV i_m is the last CAV that passes the MP ahead of CAV i. Note that if i_m and i are in the same road segment, then i_m $(= i_p)$ is the immediately preceding CAV. In this case, the safe merging constraint is redundant and need not be included.

The extended coordinator queue table $S(t)$ is updated whenever an event (as defined earlier) occurs. Thus, there are three different update processes corresponding to each triggering event:

- A new CAV enters the CZ: The CAV is indexed and added to the bottom of the queue table.
- CAV i exits the CZ: All information of CAV i is removed. All rows with index larger than i decrease their index values by 1.
- CAV i passes an MP: Mark the MP with M and update the current road segment value *curr* of CAV i with the one it is entering.

9.2.2.2 Determination of Safety Constraints

Recall that for each CAV i in the CZ, we need to consider two different safety constraints: (9.29) and (9.30). First, by looking at each row $j < i$ and the corresponding current road segment value *curr*, CAV i can determine its immediately preceding CAV i_p if one exists. This fully specifies the rear-end safety constraint (9.29).

Next, we determine the CAV which possibly conflicts with CAV i at the next MP it will pass so as to specify the safe merging constraint (9.30). To do so, we find in the extended queue table the last CAV $j < i$ that will pass or has passed the same MP as CAV i. In addition, if the CAV is in the same road segment as CAV i, it coincides with the preceding CAV i_p. Otherwise, we find i_m, if it exists. As an example, in Table 9.2 (a snapshot of Fig. 9.3), CAV 8 has no immediate preceding CAV in l_1, but it needs to yield to CAV 7 (although CAV 7 has already passed M_3, when CAV 8 arrives at M_3 there needs to be adequate space between CAV 7 and 8 for CAV 8 to enter l_4). CAV 9, however, only needs to satisfy its rear-end safety constraint with CAV 8.

It is now clear that we can use the information in $S(t)$ in a systematic way to determine both i_p in (9.29) and i_m in (9.30). Thus, there are two functions $i_p(e)$ and $i_m(e)$ which need to be updated after event e if this event affects CAV i. The index i_p can be easily determined by looking at rows $j < i$ in the extended queue table until the first one is found with the same value *curr* as CAV i. For example, CAV 9 searches for its i_p from CAV 8 to the top and sets $i_p = 8$ as CAV 8 has the *curr* value l_1. Next, the index i_m is determined. To do this, CAV i compares its MP information to that of each CAV in rows $j < i$. The process terminates the first time that any one of the following two conditions is satisfied:

- The MP information of CAV i_m *matches* CAV i. We define i_m to "match" i if and only if the last marked MP or the first unmarked MP of CAV i_m is the same as the first unmarked MP of CAV i.
- All prior rows $j < i$ have been looked up and none of them matches the MP information of CAV i.

Combining the two updating processes for i_p and i_m together, there are four different cases as follows:

1. Both i_p and i_m exist. In this case, there are two possibilities: (i) $i_p \neq i_m$. CAV i has to satisfy the safe merging constraint (9.30) with $i_p < i$ and also satisfy the rear-end safety constraint (9.29) with $i_m < i$. For example, for $i = 7$, we have $i_p = 6$ and $i_m = 4$ (M_2 is the first unmarked MP for CAV 7 and that matches the first unmarked MP for CAV 4). (ii) $i_p = i_m$. CAV i only has to follow i_p and satisfy the rear-end safety constraint (9.29) with respect to i_p. Thus, there is no safe merging constraint for CAV i to satisfy. For example, $i = 4$ and $i_p = i_m = 3$.

2. Only i_p exists. In this case, there is no safe merging constraint for CAV i to satisfy. CAV i only needs to follow the preceding CAV i_p and satisfy the rear-end safety constraint (9.29) with respect to i_p. For example, $i = 1$ and $i_p = 0$.

3. Only i_m exists. In this case, CAV i has to satisfy the safe merging constraint (9.30) with the CAV i_m in $S(t)$. There is no preceding CAV i_p, thus there is no rear-end safety constraint. For example, $i = 5$, $i_m = 1$ (M_3 is the first unmarked MP for CAV 5 and that matches the last marked MP for CAV 1 with no other match for $j = 4, 3, 2$).

4. Neither i_p nor i_m exists. In this case, CAV i does not have to consider any safety constraints. For example, $i = 2$.

9.2.2.3 Sequencing Policies with Sub-coordinator Queue Tables

Thus far, we have assumed a FIFO sequencing policy in the whole roundabout and defined a systematic process for updating $i_p(e)$ and $i_m(e)$ after each event e, hence, the entire extended queue table $S(t)$. However, FIFO may not be a good sequencing policy if applied to the whole roundabout. In order to allow possible resequencing when a CAV passes an MP, we introduce next a sub-coordinator queue table $S_k(t)$ associated with each $M_k, k = 1, 2, 3$.

Table 9.3 The sub-coordinator queue table $S_1(t)$

$S_1(t)$								
Idx	Info	Curr.	Ori.	1st MP	2nd MP	3rd MP	i_p	i_m
6	x_6	l_4	l_1	M_1, M				
7	x_7	l_4	l_1	M_1, M	M_2	M_3	6	
8	x_8	l_1	l_1	M_1	M_2			7
0	x_0	l_6	l_1	M_1, M	M_2, M	M_3, M		
1	x_1	l_6	l_1	M_1, M	M_2, M	M_3, M	0	
9	x_9	l_1	l_1	M_1	M_2	M_3	8	

$S_k(t)$ coordinates all the CAVs for which M_k is the next MP to pass or it is the last MP that they have passed. We define CZ_k as the CZ corresponding to M_k that consists of the three road segments directly connected to M_k. A sub-coordinator queue table can be viewed as a subset of the extended coordinator queue table except that the CAVs are in different order in the two tables. As an example, Table 9.3 (a snapshot of Fig. 9.3) is the sub-coordinator queue table corresponding to M_1 (in this case, still based on the FIFO policy).

The sub-coordinator queue table $S_k(t)$ is updated as follows after each event that has caused an update of the extended coordinator queue table $S(t)$:

- For each CAV i in a sub-coordinator queue table $S_k(t)$, update its information depending on the event observed: (*i*) A new CAV j enters CZ_k (either from an entry point to the roundabout CZ or a MP passing event): Add a new row to $S_k(t)$ and resequence the sub-coordinator queue table according to the sequencing policy used. (*ii*) CAV j exits CZ_k: Remove all the information of CAV j from $S_k(t)$.
- Determine i_p and i_m in each sub-coordinator queue table with the same method as described in Sect. 9.2.2.2.
- Update CAV j's i_p and i_m in the extended coordinator queue table with the corresponding information in $S_k(t)$, while M_k is the next MP of CAV j.

Note that CAV j may appear in multiple sub-coordinator queue tables with different i_p and i_m values. However, only the one in $S_k(t)$ where M_k is the next MP CAV j will pass is used to update the extended coordinator queue table $S(t)$. The information of CAV j in other sub-coordinator queue tables is necessary for determining the safety constraints as CAV j may become CAV i_p or i_m of other CAVs.

9.2.2.4 Resequencing Rule

The sub-coordinator queue table allows resequencing when a CAV passes a MP. A resequencing rule generally designs and calculates a criterion for each CAV and sorts the CAVs in the queue table when a new event happens. For example, FIFO takes the arrival time in the CZ as the criterion while the Dynamic Resequencing (DR) policy [102] uses the overall objective value in (9.33) as the criterion.

We describe here a straightforward yet effective (see Sect. 9.2.3) resequencing rule for the roundabout as follows. Let $\tilde{x}_i^k \equiv x_i - d_j^k$ be the position of CAV i relative to M_k, where d_j^k denotes the fixed distance from the entry point (origin) O_j to merging point M_k along the trajectory of CAV i. Then, consider

$$y_i(t) = -\tilde{x}_i^k(t) - \varphi v_i(t). \tag{9.34}$$

This resequencing criterion reflects the distance between the CAV and the next MP. The CAV which has the smallest $y_i(t)$ value is allocated first, thus referring to this as the Shortest Distance First (SDF) policy. Note that $\varphi v_i(t)$ introduces a speed-dependent term corresponding to the speed-dependent safety constraints. This simple resequencing rule is tested in Sect. 9.2.3. Other resequencing policies can also be easily implemented with the help of the sub-coordinator queue tables.

9.2.2.5 Unconstrained Optimal Control

We now return to the solution of problem (9.33) subject to constraints (9.27), (9.29), (9.30), (9.31), the initial condition $x_i(t_i^0) = 0$, and given t_i^0, v_i^0 and $x_i(t_i^f)$. The problem formulation is complete since we have used the extended coordinator table to determine i_p and i_m (needed for the safety constraints) associated with the closest MP to CAV i given the sequence of CAVs in the system. After introducing the sub-coordinator queue tables, we also allow some resequencing for CAVs passing each MP and focus on the CZ associated with that MP. Thus, each such problem resembles the merging control problem of Sect. 9.1 which can be analytically solved as in [83]. However, when one or more constraints become active, this solution becomes computationally intensive. The problem here is exacerbated by the fact that the values of i_p and i_m change due to different events in the roundabout system. Therefore, to ensure that a solution can be obtained in real time while also guaranteeing that all safety constraints are always satisfied, we adopt the OCBF approach from Chap. 8, which is obtained as follows: (i) an optimal control solution is first obtained for the *unconstrained* roundabout problem. (ii) This solution is used as a reference control which is optimally tracked subject to a set of CBFs, one for each of the constraints (9.29), (9.30), (9.31). Using the forward invariance property of CBFs, this ensures that these constraints are always satisfied. This whole process is carried out in a decentralized way.

With all constraints inactive (including at t_i^0), the solution of (9.33) can be derived by the same approach (the analysis is simpler due to the absence of any road curvature) as that of the merging problem of Sect. 9.1. In particular, the optimal control, velocity and position trajectories of CAV i have the form:

$$u_i^*(t) = a_i t + b_i, \tag{9.35}$$

$$v_i^*(t) = \frac{1}{2} a_i t^2 + b_i t + c_i, \tag{9.36}$$

$$x_i^*(t) = \frac{1}{6} a_i t^3 + \frac{1}{2} b_i t^2 + c_i t + d_i, \tag{9.37}$$

where the parameters a_i, b_i, c_i, d_i and t_i^f are obtained by solving a set of nonlinear algebraic equations:

$$
\begin{aligned}
&\frac{1}{2} a_i \cdot (t_i^0)^2 + b_i \cdot t_i^0 + c_i = v_i^0, \\
&\frac{1}{6} a_i \cdot (t_i^0)^3 + \frac{1}{2} b_i \cdot (t_i^0)^2 + c_i t_i^0 + d_i = 0, \\
&\frac{1}{6} a_i \cdot (t_i^f)^3 + \frac{1}{2} b_i \cdot (t_i^f)^2 + c_i t_i^f + d_i = x_i(t_i^f), \\
&a_i t_i^f + b_i = 0, \\
&\beta + \frac{1}{2} a_i^2 \cdot (t_i^f)^2 + a_i b_i t_i^f + a_i c_i = 0.
\end{aligned}
\tag{9.38}
$$

This set of equations only needs to be solved when CAV i enters the CZ and, as already mentioned, it can be done very efficiently.

9.2.2.6 Joint Optimal Control and Barrier Function (OCBF) Method

Once we obtain the unconstrained optimal control solution (9.35) through (9.37), we use a function $h(u_i^*(t), x_i^*(t), x_i(t))$ as a control reference $u_{ref}(t) = h(u_i^*(t), x_i^*(t), x_i(t))$, where $x_i(t)$ provides feedback from the actual observed CAV trajectory. We then design a controller by proceeding exactly as in Sect. 9.1.2.4.

First, let $\boldsymbol{x}_i(t) \equiv (x_i(t), v_i(t))$. Due to the vehicle dynamics (9.27), define $f(\boldsymbol{x}_i(t)) = [v_i(t), 0]^T$ and $g(\boldsymbol{x}_i(t)) = [0, 1]^T$. The constraints (9.29), (9.30) and (9.31) are expressed in the form $b_k(\boldsymbol{x}_i(t)) \geq 0, k \in \{1, \ldots, b\}$ where b is the number of constraints. The CBF method maps the constraint $b_k(\boldsymbol{x}_i(t)) \geq 0$ to a new constraint, which directly involves the control $u_i(t)$ and takes the usual form

$$L_f b_k(\boldsymbol{x}_i(t)) + L_g b_k(\boldsymbol{x}_i(t)) u_i(t) + \gamma(b_k(\boldsymbol{x}_i(t))) \geq 0. \tag{9.39}$$

To optimally track the reference speed trajectory, a CLF function $V(\boldsymbol{x}_i(t))$ is used. Letting $V(\boldsymbol{x}_i(t)) = (v_i(t) - v_{ref}(t))^2$, the CLF constraint takes the form

$$L_f V(\boldsymbol{x}_i(t)) + L_g V(\boldsymbol{x}_i(t)) u_i(t) + \epsilon V(\boldsymbol{x}_i(t)) \leq e_i(t), \tag{9.40}$$

where $\epsilon > 0$, and $e_i(t)$ is a relaxation variable which makes the constraint soft. Then, the OCBF controller optimally tracks the reference trajectory by solving the optimization problem:

$$\min_{u_i(t),e_i(t)} \int_{t_i^0}^{t_i^f} \left(\beta e_i^2(t) + \frac{1}{2}(u_i(t) - u_{ref(t)})^2 \right) \tag{9.41}$$

subject to the vehicle dynamics (9.27), the CBF constraints (9.39) and the CLF constraints (9.40). As already discussed in Chap. 8, there are several possible choices for $u_{ref}(t)$ and $v_{ref}(t)$. In the sequel, we choose the simplest and most straightforward one:

$$v_{ref}(t) = v_i^*(t), \quad u_{ref}(t) = u_i^*(t), \tag{9.42}$$

where $v_i^*(t)$ and $u_i^*(t)$ are obtained from (9.36) and (9.35).

Considering all constraints in (9.33), the rear-end safety constraint (9.29) and the vehicle limitations (9.31) are straightforward to transform into a CBF form. As an example, consider (9.29) by setting $b_1(x_i(t)) = z_{i,i_p}(t) - \varphi v_i(t) - \delta$. After calculating the Lie derivatives, the CBF constraint (9.39) can be obtained. The safe merging constraint (9.30) differs from the rest in that it only applies to specific time instants t_i^{mk}. This poses a technical complication due to the fact that a CBF must always be in a continuously differentiable form. We can convert (9.30) to such a form using the technique in [73] to obtain

$$z_{i,i_m}(t) - \Phi(x_i(t))v_i(t) - \delta \geq 0, \quad t \in [t_i^{k,0}, t_i^k] \tag{9.43}$$

where $t_i^{mk,0}$ denotes the time CAV i enters the road segment connected to M_k and $\Phi(\cdot)$ is any strictly increasing function as long as it satisfies the boundary constraints $z_{i,i_m}(t_i^{k,0}) - \phi v_i(t_i^{k,0}) - \delta \geq 0$ and $z_{i,i_m}(t_i^k) - \phi v_i(t_i^k) - \delta \geq 0$ (which is precisely (9.30)). Note that we need to satisfy (9.43) when a CAV changes road segments in the roundabout and the value of i_m changes. Since $z_{i,i_m}(t_i^{k,0}) \geq -L_{i_m} + L_i$, where L_i is the length of the road segment CAV i is in, to guarantee the feasibility of (9.43), we set $\Phi(x_i(t_i^{k,0}))v_i(t_i^{k,0}) + \delta = -L_{i_m} + L_i$. Then, from (9.30), we get $\Phi(x_i(t_i^k)) = \varphi$. Simply choosing a linear $\Phi(\cdot)$ as follows:

$$\Phi(x_i(t)) = \left(\varphi + \frac{L_{i_m} - L_i + \delta}{v_i(t_i^{k,0})} \right) \frac{x_i(t)}{L_i} - \frac{L_{i_m} - L_i + \delta}{v_i(t_i^{k,0})} \tag{9.44}$$

it is easy to check that it satisfies the boundary requirements. Note that when implementing the OCBF controller, $x_i(t)$ needs to be transformed into a relative position $\tilde{x}_i^k + L_i$, which reflects the distance between CAV i and the origin of the current road segment. Then, z_{i,i_p} and z_{i,i_m} are calculated after this transformation, where $z_{i,i_p} = \tilde{x}_{i_p}^k - \tilde{x}_i^k$, $z_{i,i_p} = \tilde{x}_{i_m}^k - \tilde{x}_i^k$.

With all constraints converted to CBF constraints in (9.41), the solution is obtained, as in the traffic merging control problem, through the familiar method first presented at the end of Chap. 2 : we discretize the time interval $[t_i^0, t_i^m]$ with time steps of length Δ and solve (9.41) over $[t_i^0 + k\Delta, t_i^0 + (k+1)\Delta]$, $k = 0, 1, \ldots$, with $u_i(t), e_i(t)$ as decision variables held constant over each such interval. Consequently, each such problem is a QP since we have a quadratic cost and a number of linear constraints on the decision variables at the beginning of each interval. The solution of each such problem gives an optimal control $u_i^*(t_i^0 + k\Delta)$,

$k = 0, 1, \ldots$, allowing us to update the vehicle dynamic in the kth time interval. This process is repeated until CAV i leaves the CZ.

9.2.3 Simulation Results for Traffic Control in Roundabouts

As in Sect. 9.1.3, we use the multi-model traffic flow simulation platform Vissim, as a baseline to compare the performance of traffic consisting of all CAVs in a roundabout as shown in Fig. 9.3 using the OCBF controller to that of human-driven vehicles as a baseline.

Simulation 1: The first simulation focuses on the performance of the OCBF controller. The basic parameter settings are as follows: $L_a = 60\,\text{m}$, $L = 60\,\text{m}$, $\delta = 10\,\text{m}$, $\varphi = 1.8\,\text{s}$, $v_{max} = 17\,\text{m/s}$, $v_{min} = 0$, $u_{max} = 5\,\text{m/s}^2$, $u_{min} = -5\,\text{m/s}^2$. This scenario considers a *symmetric* configuration in the sense that $L_a = L$. The traffic in the three incoming roads is generated through Poisson processes with all rates set to 360 CAVs/h. Under these traffic rates, vehicles will sometimes form queues waiting for other vehicles to go through the roundabout. A total number of approximately 200 CAVs are simulated. The simulation results of the performance of OCBF compared to that in Vissim are listed in Table 9.4.

In this simulation, FIFO is chosen as the sequencing policy in the OCBF method. As seen in Table 9.4, the travel time of CAVs in the roundabout improves by about 34% using the OCBF method compared with that of Vissim when $\alpha = 0.1$ (with some additional improvement when $\alpha = 0.2$). The CAVs using the OCBF method consume 52 and 26% less energy than that in Vissim with α set to 0.1 and 0.2 respectively. A larger α value means more emphasis on the travel time than the energy consumption, which explains the shorter travel time and the larger energy consumption. When it comes to the total objective, the OCBF controller shows 44 and 32% improvement over the human-driven performance in Vissim when α equals to 0.1 and 0.2 respectively. This improvement in both the travel time and the energy consumption is to be expected as the CAVs using the OCBF method never stop and wait for CAVs in another road segment to pass as in Vissim.

Table 9.4 Objective function comparison for a symmetric roundabout

Items	OCBF		Vissim	
Weight	$\alpha = 0.1$	$\alpha = 0.2$	$\alpha = 0.1$	$\alpha = 0.2$
Ave. time (s)	13.7067	13.3816	20.6772	
Ave. energy	16.0698	24.6336	33.2687	
Ave. obj. [1]	35.1084	66.4511	61.9893	97.8850

[1] Ave. obj $= \beta \times$ Ave. time + Ave. energy,
$\beta = \dfrac{\alpha \max\{u_{max}^2, u_{min}^2\}}{2(1-\alpha)}$

Table 9.5 Objective function comparison of different resequencing rules in an asymmetric roundabout

Items	OCBF+FIFO	OCBF+SDF	Vissim
Ave. time (s)	16.4254	14.7927	24.6429
Ave. energy	56.9643	23.1131	30.8947
Ave. obj.	108.2937	69.3403	107.9038

Simulation 2: The second simulation compares the performance of OCBF under different sequencing rules in an *asymmetric* configuration. The parameter settings are the same as the first case except that $L = 100$ m. The weight is set to $\alpha = 0.2$. The simulation results of the performance of OCBF with FIFO and OCBF with the SDF sequencing policy, as well that in Vissim, are shown in Table 9.5.

Table 9.5 shows that a CAV using OCBF with FIFO spends around 33% less travel time but 84% more energy than that in Vissim. The average objective values of the two cases are almost the same, indicating that OCBF with FIFO works poorly in an asymmetric roundabout. For example, when a CAV enters segment l_4, it has to wait for another CAV that has entered l_2 just before it to run 40 more meters for safe merging. This is unreasonable and may also result in some extreme cases when the OCBF problem becomes infeasible. This problem can be resolved by choosing a better sequencing policy such as SDF. As shown in Table 9.5, OCBF+SDF outperforms OCBF+FIFO, achieving an improvement of 40% in travel time, 26% in energy consumption and 36% in the objective value compared to that in Vissim.

Simulation 3: The purpose of this simulation experiment is to study the effect of traffic volume. The roundabout is set to be *asymmetric* with the same parameter settings as in **Simulation 2**. A total number of approximately 500 CAVs are simulated under two sets of incoming traffic rates: balanced incoming traffic (360 CAVs/h for each origin) and imbalanced incoming traffic (540 CAVs/h from O_1, 270 CAVs/h from O_2 and O_3). The simulation results of the performance of OCBF+SDF compared to that in Vissim under both balanced and imbalanced incoming traffic are shown in Tables 9.6 and 9.7.

From Table 9.6, it is seen that the imbalanced traffic causes longer travel times ($\sim$2s) and more energy consumption ($\sim$13%) although the total traffic rates are the same. The imbalanced traffic results in an imbalanced performance of CAVs from different origins. The CAVs originated from O_1 with heavy traffic perform worse than those from O_2 and O_3 with light traffic. However, when OCBF+SDF is applied to the system, the imbalanced traffic brings almost no performance loss and becomes somewhat more balanced after passing the roundabout. This result is interesting because the fact that traffic is *imbalanced* was not taken into consideration in the OCBF method. An explanation of this phenomenon is that SDF gives the CAVs from O_1 a higher priority as they are more likely to be the closest to the

Table 9.6 Objective function comparison (Vissim)

Traffic type	CAV Origin	Time	Energy	Ave. Obj.
Balanced	All	24.7615	32.1295	109.5092
	From O_1	24.7631	33.7724	111.1570
	From O_2	23.6934	31.4432	105.4851
	From O_3	25.8458	31.0375	111.8057
Imbalanced	Total	26.5170	36.0547	118.9203
	From O_1	28.1765	39.5067	127.5582
	From O_2	25.6977	35.5803	115.8857
	From O_3	23.8207	29.2346	103.6744

Table 9.7 Objective function comparison (OCBF+SDF)

Traffic type	CAV Origin	Time	Energy	Ave. Obj.
Balanced	All	15.1650	24.8601	72.2507
	From O_1	15.1613	24.8453	72.2245
	From O_2	14.8221	25.3610	71.6801
	From O_3	15.5156	24.3696	72.8560
Imbalanced	Total	15.2437	24.1693	71.8059
	From O_1	15.4035	24.9519	73.0880
	From O_2	14.6235	19.8901	65.5885
	From O_3	15.5163	26.7188	75.2072

MP while OCBF allows the CAVs to pass the roundabout quickly without stop; therefore, the CAVs from a heavy traffic flow are less likely to get congested.

9.3   Multi-lane Signal-Free Intersections

Intersections are the main bottlenecks in urban traffic networks. As reported in [52], congestion in these areas causes US commuters to spend 6.9 billion hours more on the road and to purchase an extra 3.1 billion gallons of fuel, resulting in a substantial economic loss to society. The coordination and control problems at intersections are particularly challenging in terms of safety, traffic efficiency, and energy consumption [15, 52].

In this section, as in the last two, we also combine the use of Control Barrier Functions (CBFs) with conventional optimal control to seek optimality in the performance of CAVs while provably guaranteeing the on-line safe execution of their trajectories. The key idea (see [94]) is to design CAV trajectories which optimally track analytically tractable solutions of the basic intersection-crossing optimization problem while also ensuring that all safety constraints are satisfied. As in the cases of merging and roundabout traffic control, we use CBFs to map continuously differentiable state constraints into new control-based constraints using the techniques developed through the book, starting with Chap. 2. Due to the forward invariance of the associated safe set, a control input that satisfies these new constraints is also guaranteed to satisfy the original state constraints. This property makes the CBF method effective even when the vehicle dynamics and constraints become complicated and include noise.

9.3.1 Multi-lane Signal-Free Intersection Traffic Control as an Optimal Control Problem

Figure 9.4 shows a typical intersection with multiple lanes. The area within the outer red circle is called the *Control Zone* (CZ), and the length of each CZ segment is L_1 which is initially assumed to be the same for all entry points to the intersection; extensions to asymmetrical intersections are straightforward and discussed later in this section. The significance of the CZ is that it allows all vehicles to share information and be automatically controlled while in it, as long as all vehicles in the CZ are CAVs. Red dots show all Merging Points (MPs) where potential collisions may occur. We assume that the motion trajectory of each CAV in the intersection is determined upon its entrance to the CZ (see grey lines in Fig. 9.4). Based on these trajectories, all MPs in the intersection are fixed and can be easily determined. However, unlike most prior similar studies, we also allow possible lane-changing behaviors in the CZ, which adds generality to this analysis. In order to avoid potential collisions with newly arriving vehicles and to conform to the driving rules that vehicles are prohibited from changing lanes when they are very close to the intersection, these lane changes are only allowed in a "lane-changing zone," i.e., an area between the two blue lines shown in Fig. 9.4. We use dotted lines to differentiate these zones from the rest of the area where solid lines are shown. The distance from the entry of the CZ to the lane-changing zone is L_2 and the length of the lane-changing zone is L_3. Since we initially consider a symmetrical intersection, L_2 and L_3 are the same for all lanes. However, it is easy to extend this approach to asymmetrical intersections and set different parameters for each lane, as shown in the sequel. Due to lane-changing, apart from the fixed MPs in the intersection, some "floating" MPs may also appear in lane-changing zones which are not fixed in space. Thus, there are two kinds of MPs: (i) the fixed MPs in the intersection and (ii) the floating MPs in the lane-changing zones.

We label the lanes from l_1 to l_8 in a counterclockwise direction with corresponding origins O_1 to O_8. The rightmost lanes in each direction only allow turning right or going straight, while the leftmost lanes only allow turning left or going straight. However, all CAVs have three possible movements: going straight, turning left, and turning right. Thus, some CAVs must change their lanes so as to execute a movement, e.g., left-turning CAV 2 in l_2 in Fig. 9.4. Due to such lane-changing behavior, a new MP $M_{2,1}$ is generated since a conflict

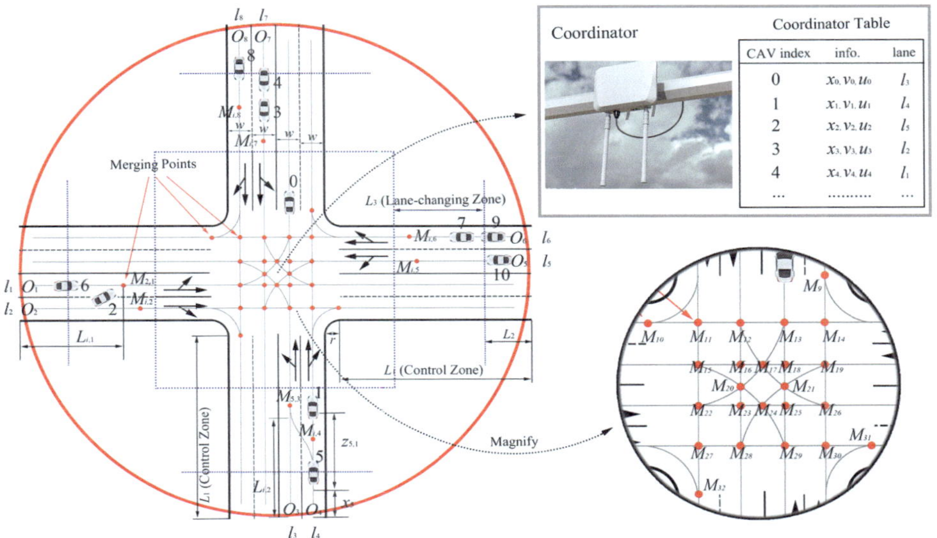

Fig. 9.4 The multi-lane signal-free intersection problem. Collisions may happen at the MPs (red dots shown in above figure)

of CAV 2 with a CAV in l_1 may arise. Similarly, possible MPs may also appear in other lanes when vehicles perform lane-changing maneuvers, as the red dots ($M_{i,2}$, $M_{i,3}$, ..., and $M_{i,8}$) indicate in Fig. 9.4. Moreover, it is worth noting that if a CAV needs to change lanes, then it has to travel an additional (lateral) distance; we assume that this extra distance is a constant $l > 0$. In what follows, we consider an intersection that has two lanes in each direction, which, in our view, represents one of the most common intersection configurations worldwide, and observe that this model is easy to generalize to intersections with more than two lanes.

The intersection has a coordinator (typically a Road Side Unit (RSU)) whose function is to maintain the crossing sequence and all individual CAV information. As already discussed in prior sections, the most common crossing sequence is based on the First-In-First-Out (FIFO) queue formed by all CAVs using their arrival time at the CZ, regardless of the lane each CAV belongs to. The FIFO queue is fair and simple to implement, however, its performance can occasionally be poor, especially when an intersection is *asymmetric* in terms of its geometry or the traffic rates at different entry points. Thus, various cooperative driving strategies have been proposed to generate a more promising crossing sequence, as in [82, 96, 102]. Our approach for controlling CAVs does not depend on the specific crossing sequence selected. Therefore, we first use the FIFO queue so as to enable accurate comparisons with related work which also uses this scheme and then generalize it to include other resequencing methods which adjust the crossing sequence whenever there is a new arriving vehicle, e.g., the Dynamic Resequencing (DR) method in [102]. This allows CAVs to overtake other CAVs in the CZ from different roads, which better captures actual intersection traffic behaviors.

The signal-free intersection model described so far introduces several new elements to prior intersection models, including: (i) replacing merging *zones* with merging *points*, which are less conservative since they reduce the spacing between CAVs; (ii) allowing lane-changing; (iii) allowing resequencing to improve the intersection throughput. In subsequent sections, the following additional features will be introduced: (iv) an extended coordinator table to identify pairs of conflicting CAVs at MPs, necessary to enforce safe merging constraints; (v) safety guarantees provided through the use of CBFs; (vi) we allow noise and more complicated vehicle dynamics through the use of the OCBF controller.

When a CAV enters the CZ, the coordinator will assign it a unique index. Let $S(t)$ be the set of FIFO-ordered CAV indices and $N(t)$ be the cardinality of $S(t)$. Based on $S(t)$, the coordinator stores and maintains an information table, as shown in Fig. 9.4. For example, the current lane of CAV 2 changes from l_2 to l_1 after it completes a lane-changing maneuver. In addition, after CAV 0 passes the intersection, its index will be dropped from the table and the indices of all other CAVs decrease by one. This table enables each CAV to quickly identify other CAVs that have potential collisions with it and to optimize its trajectory to maximize some specific objectives. The search algorithm for identifying conflicting vehicles will be introduced in detail in the next section.

The vehicle dynamics for CAV i take the form

$$\dot{x}_i(t) = v_i(t) + w_{i,1}(t), \qquad \dot{v}_i(t) = u_i(t) + w_{i,2}(t), \tag{9.45}$$

where $x_i(t)$ is the distance to its origin along the lane that CAV i is located in when it enters the CZ, $v_i(t)$ denotes the velocity, and $u_i(t)$ denotes the control input (acceleration). Moreover, to compensate for possible measurement noise and modeling inaccuracy, we use $w_{i,1}(t)$ and $w_{i,2}(t)$ to denote two random processes defined in an appropriate probability space.

We can now formulate an optimization problem where we consider two objectives for each CAV subject to three constraints, including the rear-end safety constraint with the preceding vehicle in the same lane, the lateral safety constraints with vehicles in the other lanes, and the vehicle physical constraints, as detailed next.

Objective 1 (*Minimize travel time*): Let t_i^0 and t_i^m denote the time that CAV i arrives at the origin and the time that it enters the intersection, respectively. To improve traffic efficiency, we wish to minimize the travel time $t_i^m - t_i^0$ for CAV i.

Objective 2 (*Minimize energy consumption*): Apart from traffic efficiency, another objective is energy efficiency. Ignoring any noise terms in (9.45) for the time being, since the energy consumption rate is a monotonic function of the acceleration control input, an energy consumption function we use is defined as

$$J_i = \int_{t_i^0}^{t_i^m} \mathcal{C}(u_i(t))dt, \tag{9.46}$$

where $\mathcal{C}(\cdot)$ is a strictly increasing function of its argument.

Constraint 1 (*Rear-end safety constraint*): Let i_p denote the index of the CAV which physically immediately precedes i in the CZ (if one is present). To avoid rear-end collisions, we require that the spacing $z_{i,i_p}(t) \equiv x_{i_p}(t) - x_i(t)$ be constrained by:

$$z_{i,i_p}(t) \geq \varphi v_i(t) + \delta, \quad \forall t \in [t_i^0, t_i^m], \tag{9.47}$$

where δ is the minimum safety distance, and φ denotes the reaction time (as a rule, $\varphi = 1.8\,\mathrm{s}$ is suggested, e.g., [67]). If we define z_{i,i_p} to be the distance from the center of CAV i to the center of CAV i_p, then δ is a constant determined by the length of these two CAVs (thus, δ is generally dependent on CAVs i and i_p but taken to be a constant over all CAVs in the sequel, only for simplicity). Note that i_p may change when a lane change event or an overtaking event occurs.

Constraint 2 (*Lateral safety constraint*): Let t_i^k denote the time that CAV i arrives at the MP M_k, $k \in \{1, 2, \ldots, 32\}$. CAV i may collide with other vehicles that travel through the same MP. For all MPs, including the floating MPs $M_{i,l}$ due to lane-changing, there must be enough safe space when CAV i is passing through, that is,

$$z_{i,j}(t_i^k) \geq \varphi v_i(t_i^k) + \delta, \tag{9.48}$$

where $j \neq i$ is a CAV that may collide with i (note that j may not exist and that there may also be multiple CAVs indexed by j for which this constraint applies at different t_i^k). The determination of j is challenging, and will be addressed in the following section. Compared with related work that requires no more than one CAV within a conflict (or merging) zone at any time, we use (9.48) to replace this conservative constraint. Instead of such a fixed zone, the space defining collision avoidance around each MP now depends on the CAV's speed (and possibly size if we allow δ to be CAV-dependent), hence it is much more flexible.

Constraint 3 (*Vehicle physical limitations*): Due to the physical limitations on motors and actuators, there are physical constraints on the velocity and control inputs for each CAV i:

$$\begin{aligned}
v_{\min} &\leq v_i(t) \leq v_{\max}, \quad \forall t \in [t_i^0, t_i^m], \\
u_{i,\min} &\leq u_i(t) \leq u_{i,\max}, \quad \forall t \in [t_i^0, t_i^m],
\end{aligned} \tag{9.49}$$

where $v_{\max} > 0$ and $v_{\min} \geq 0$ denote the maximum and minimum velocity allowed in the CZ, while $u_{i,\min} < 0$ and $u_{i,\max} > 0$ denote the minimum and maximum control input for each CAV i, respectively. We assume that all vehicles are homogeneous passenger cars and their minimum and maximum control inputs are the same. Thus, in what follows, we use $u_{\min}$ and $u_{\max}$ instead of $u_{i,\min}$ and $u_{i,\max}$; this is done only for simplicity and does not limit our analysis.

We use a quadratic function for $\mathcal{C}(u_i(t))$ in (9.46) thus minimizing energy consumption by minimizing the control input effort $\frac{1}{2}u_i^2(t)$. By normalizing travel time and $\frac{1}{2}u_i^2(t)$, and using $\alpha \in [0, 1)$, we construct a convex combination as follows:

$$\min{}_{u_i(t),t_i^m} \int_{t_i^0}^{t_i^m} \left(\alpha + \frac{(1-\alpha)\frac{1}{2}u_i^2(t)}{\frac{1}{2}\max\{u_{\max}^2, u_{\min}^2\}} \right) dt. \tag{9.50}$$

If $\alpha = 1$, problem (9.50) is equivalent to a minimum travel time problem; if $\alpha = 0$, it becomes a minimum energy consumption problem.

By defining $\beta \equiv \frac{\alpha\max\{u_{\max}^2, u_{\min}^2\}}{2(1-\alpha)}$ (assuming $\alpha < 1$) and multiplying (9.50) by the constant $\frac{\beta}{\alpha}$, we have:

$$\min{}_{u_i(t),t_i^m} \beta(t_i^m - t_i^0) + \int_{t_i^0}^{t_i^m} \frac{1}{2}u_i^2(t)dt, \tag{9.51}$$

where $\beta \geq 0$ is a weight factor that can be adjusted through $\alpha \in [0, 1)$ to penalize travel time relative to the energy cost. This minimization problem is subject to constraints (9.45), (9.47), (9.48), (9.49), given t_i^0 and the initial and final conditions $x_i(t_i^0) = 0$, $v_i(t_i^0)$, $x_i(t_i^m)$.

9.3.2   Decentralized Online Control Framework

The solution of problem (9.51) can be obtained as described in Sect. 9.1.2 (see also [83]) where a single MP is involved in a two-road single-lane merging configuration where the value of j in (9.5) is immediately known. The difficulty here is that there may be more than one CAV j defining lateral safety constraints for any $i \in S(t)$. Determining the value(s) of j is challenging since there are eight lanes and three possible movements at intersections as shown in Fig. 9.4. Obviously, this can become even harder as more lanes are added or asymmetrical intersections are considered. Therefore, we propose a general MP-based approach which involves two steps. The first step addresses the following two issues: (i) a strategy for determining "floating" MPs due to CAVs possibly changing lanes within the CZ, and (ii) a strategy for determining all lateral safety constraints, hence the values of j in (9.5). Once these issues are addressed in the following subsections, problem (9.51) is well-defined. The second step consists of solving problem (9.51) through an OCBF controller as described in Chap. 8 and similar to the solutions of the merging and roundabout traffic control problems in prior sections. The overall process is outlined in **Algorithm** 1 which is implemented in time-driven manner by replanning the control inputs of all CAVs every T seconds.

Algorithm 1: MP-based Algorithm for Multi-lane intersection problems

---

Initialize an empty queue table $S(t)$;

for *every T seconds* **do**

    if *a new vehicle enters the CZ* **then**

        Determine a passing order for all CAVs according to the FIFO rule or other resequencing methods, e.g., the DR scheme;

        Plan an *unconstrained* optimal control trajectory for the new CAV;

        if *the new CAV needs to change lanes* **then**

            Use the lane-changing MP determination strategy to determine the lane-changing location and time for the new CAV;

        end

        Add the information of the new CAV into $S(t)$;

    end

    for *each CAV in $S(t)$* **do**

        Use the lateral safety constraint determination strategy to determine which constraints it needs to meet;

        Use the OCBF controller to obtain control inputs for it;

        if *this CAV has left the intersection* **then**

            Remove the information of this CAV from $S(t)$;

        end

    end

end

9.3.2.1 Determination of Lane-Changing MPs

When a new CAV $i \in S(t)$ arrives at the origins O_2, O_4, O_6, O_8 (or O_1, O_3, O_5, O_7) and must turn left (or right), it has to change lanes before getting close to the intersection. Therefore, CAV i must determine the location of the variable (floating) MP $M_{i,k}$, $k \in \{1, 2, \ldots, 8\}$.

There are three important observations to make:

(i) The *unconstrained* optimal control for such i is independent of the location of $M_{i,k}$, $k \in \{1, 2, \ldots, 8\}$ since we have assumed that lane-changing will only induce a fixed extra length l regardless of where it occurs.

(ii) The optimal control solution under the lateral safety constraint is better (i.e., lower cost in (9.51)) than one which includes an active rear-end safety constrained arc in its optimal trajectory. This is because the former applies only to a single time instant t_i^k whereas the latter requires the constraint (9.47) to be satisfied over all $t \in [t_i^0, t_i^k]$. It follows that any MP $M_{i,k}$ should be as close as possible to the intersection (i.e., $L_{i,k}$ should be as large as possible, and its maximum value is $L_2 + L_3$), since the lateral safety constraint after $M_{i,k}$ will become a rear-end safety constraint with respect to some j in the adjacent lane. For instance, suppose that CAV 10 in Fig. 9.4 is a right-turn vehicle. After changing lanes to the destination lane l_6, it will be constrained by the rear-end safety constraint with its preceding vehicle (e.g., CAV 9), which will influence its optimal trajectory. Thus, to reduce the impact time, CAV 10 should change lanes as late as possible.

(iii) In addition, CAV i may also be constrained by its physically preceding CAV i_p (if one exists) in the same lane as i. In this case, CAV i needs to consider both the rear-end safety constraint with i_p and the lateral safety constraint with some $j \neq i$. Thus, the solution is more constrained (hence, more sub-optimal) if i stays in the current lane after the rear-end safety constraint due to i_p becoming active. We conclude that in this case CAV i should change its lane to the left (right) lane as late as possible, i.e., just as the rear-end safety constraint with i_p first becomes active, i.e., $L_{i,k}$ is determined by

$$L_{i,k} = x_i^*(t_i^a), \qquad (9.52)$$

where $x_i^*(t)$ denotes the unconstrained optimal trajectory of CAV i, and $t_i^a \geq t_i^0$ is the time instant when the rear-end safety constraint between i and i_p first becomes active; if this constraint never becomes active, then $L_{i,k} = L_2 + L_3$. The value of t_i^a is determined from (9.47) by

$$x_{i_p}^*(t_i^a) - x_i^*(t_i^a) = \varphi v_i^*(t_i^a) + \delta, \qquad (9.53)$$

where $x_{i_p}^*(t)$ is the *unconstrained* optimal position of CAV i_p and $v_i^*(t)$ is the *unconstrained* optimal speed of CAV i. If, however, CAV i_p's optimal trajectory itself happened to include a constrained arc, then (9.53) is only an upper bound of t_i^a.

In summary, it follows from (i) $-$ (iii) above that if CAV i never encounters a point in its current lane where its rear-end safety constraint becomes active, we set $L_{i,k} = L_2 + L_3$, otherwise, $L_{i,k}$ is determined through (9.52)–(9.53).

We note that the distances from the origins $O_1, \ldots, O_8$ to MPs are not all the same, and the CAVs that make a lane change will induce an extra l distance. Therefore, we need to perform a coordinate transformation for those CAVs that are in different lanes and will meet at the same MP M_k, $k \in \{1, \ldots, 32\}$. In other words, when $i \in S(t)$ obtains information for $j \in S(t)$ from the FIFO queue table to account for the lateral safety constraint at an MP M_k, the position information $x_j(t)$ is transformed into $x_j'(t)$ through

$$x_j'(t) := \begin{cases} x_j(t) + L_{i,k} - L_{j,k} + l, & \text{if only } i \text{ changes lane,} \\ x_j(t) + L_{i,k} - L_{j,k} - l, & \text{if only } j \text{ changes lane,} \\ x_j(t) + L_{i,k} - L_{j,k}, & \text{otherwise} \end{cases} \qquad (9.54)$$

where $L_{i,k}$ and $L_{j,k}$ denote the distances of the MPs M_k from the origins of CAVs i and j, respectively. Note that the coordinate transformation (9.54) only applies to CAV i obtaining information on j from $S(t)$, and does not involve any action by the coordinator.

9.3.2.2 Determination of Lateral Safety Constraints

We begin with the observation (by simple inspection of Fig. 9.4) that CAVs can be classified into two categories, depending on the lane that a CAV arrives at, as follows:

Table 9.8 The extended coordinator queue table for Fig. 9.4

$S_1(t)$

Idx	Info.	Curr.	Ori.	1st MP	2nd MP	3rd MP	4th MP	5th MP
0	x_0, v_0, u_0	l_3	l_3	M_{29}	M_{25}	M_{21}	M_{18}	M_{13}
1	x_1, v_1, u_1	l_4	l_4	M_{30}	M_{26}	M_{19}	M_{14}	M_9
2	x_2, v_2, u_2	l_2	l_2	$M_{2,1}$	M_{22}	M_{20}	M_{17}	M_{13}
3	x_3, v_3, u_3	l_7	l_7	M_{12}	M_{16}	M_{20}	M_{23}	M_{28}
4	x_4, v_4, u_4	l_7	l_7	M_{12}	M_{16}	M_{20}	M_{23}	M_{28}
5	x_5, v_5, u_5	l_4	l_4	$M_{5,3}$	M_{29}	M_{24}	M_{20}	M_{15}
6	x_6, v_6, u_6	l_1	l_1	M_{22}	M_{20}	M_{17}	M_{13}	
7	x_7, v_7, u_7	l_6	l_6	M_9				
8	x_8, v_8, u_8	l_8	l_8	M_{10}				
9	x_9, v_9, u_9	l_6	l_6	M_{14}	M_{13}	M_{12}	M_{11}	M_{10}
10	x_{10}, v_{10}, u_{10}	l_5	l_5	M_{19}	M_{18}	M_{17}	M_{16}	M_{15}

1. The CAVs arriving at lanes l_1, l_3, l_5, l_7 will pass

   - two MPs if they choose to turn right (including the floating MP $M_{i,k}$, $k \in \{2, 4, 6, 8\}$);
   - four MPs it they turn left;
   - five MPs if they go straight.

2. The CAVs arriving at lanes l_2, l_4, l_6, l_8 will pass

   - only one MP if they choose to turn right;
   - five MPs if they turn left (including the floating MP $M_{i,k}$, $k \in \{1, 3, 5, 7\}$) or if they go straight.

Clearly, the maximum number of MPs a CAV may pass is 5. Since all such MPs are determined upon arrival at the CZ, we augment the queue table in Fig. 9.4 by adding the original lane and the MP information for each CAV as shown in Table 9.8 for a snapshot of Fig. 9.4. This is an extended coordinator queue table similar to the one introduced in Sect. 9.2.2.1. The current and original lanes are shown in the third and fourth column, respectively. The original lane is fixed, while the current lane may vary dynamically: in **Algorithm** 1, the state of all CAVs in the queue is updated if any of them has changed lanes. The remaining five columns show all MPs a CAV will pass through in order. For example, left-turning CAV 2 in Fig. 9.4 passes through five MPs $M_{2,1}$, M_{22}, M_{20}, M_{17}, and M_{13} sequentially, where we label the 1st MP as $M_{2,1}$ and so forth.

When a new CAV enters the CZ, it first determines whether it will change lanes (as described above) and identifies all MPs that it must pass. At this point, it looks at the extended queue table $S(t)$ (an example is shown in Table 9.8) which already contains all prior CAV states and MP information. First, from the current lane column, CAV i can determine its current physically immediately preceding CAV i_p if one exists. Next, since the passing priority has been determined by the sequencing method selected (FIFO or otherwise), CAVs need to yield to other CAVs that rank higher in the queue $S(t)$. In addition, for any MP that CAV i will pass through, it only needs to yield to the closest CAV that has higher priority than it, and this priority is determined by the order of $S(t)$. For instance, CAVs 3, 4, and 5 will all pass through M_{20}, as shown in Table 9.8. For the MP M_{20}, CAV 5 only needs to meet the lateral safety constraint with CAV 4, while the constraint with CAV 3 will be automatically met since CAV 4 yields to CAV 3. Similarly, we can find indices of CAVs for other MPs crossed by CAV 5. Since CAV 5 passes through five MPs, we define an index set Ω_5 for CAV 5 which has at most 5 elements. In this example, CAV 5 only conflicts with CAV 0 at M_{29} besides M_{20}, so that $\Omega_5 = \{0, 4\}$.

Algorithm 2: Search Algorithm for Conflict CAVs

Input: The extended coordinator queue table $S(t)$, CAV i
Output: The index of conflict CAVs for CAV i
Initialize an empty index set Ω_i;
Initialize a set Θ_i including all MPs CAV i will pass through;
Find the position k of CAV i in the $S(t)$;
for $j = k - 1 : -1 : 1$ **do**
    if *the jth CAV in $S(t)$ passes at least one MP in Θ_i* **then**
        Add the index of this CAV into the set Ω_i;
        Remove the same MPs from Θ_i;
    end
    if Θ_i *is empty* **then**
        break;
    end
end
return Ω_i

Therefore, it remains to use the information in $S(t)$ in a systematic way so as to determine all the indices of those CAVs that CAV i needs to yield to; these define each index j in (9.48) constituting all lateral safety constraints that CAV i needs to satisfy. This is accomplished by a search algorithm (**Algorithm** 2) based on the following process. CAV i compares its original lane and MP information to that of every CAV in the table starting with the last row of itself and moving up. The process terminates the first time that any one of the following three conditions is satisfied at some row $j < i$:

1. All MP information of CAV i matches row j and Ω_i is empty.
2. Every MP for CAV i has been matched to some row j.
3. All prior rows $j < i$ have been looked up.

These three conditions are examined in order:

Constraint 1   If this is satisfied, there are no conflicting MPs for CAV i and this implies that CAV i_p is the physically immediately preceding CAV all the way through the CZ. Thus, CAV i only has to satisfy the safety constraint (9.47) with respect to i_p, i.e., it just follows CAV i_p. For example, $i = 4$ and $i_p = 3$ in Fig. 9.4 (and Table 9.8).

Constraint 2   If this holds, then CAV i has to satisfy several lateral safety constraints with one or more CAV $j \in \Omega_i$. Moreover, it also has to satisfy the rear-end safety constraint (9.47) with CAV i_p, where i_p is determined by the first matched row in the current lane column of Table 9.8. For example, $i = 10$, $j = 0, 1, 4, 5,$ and 6 in Fig. 9.4 (and Table 9.8).

Constraint 3   There are two cases. First, if the index set Ω_i is empty, then CAV i does not have to satisfy any lateral safety constraint; for example, $i = 7$ in Fig. 9.4 (and Table 9.8). Otherwise, it needs to yield to all CAVs in Ω_i; for example, $i = 2$, $j = 0$ in Fig. 9.4 (and Table 9.8).

We observe that **Algorithm** 2 can be implemented for all CAVs in an event-driven way (since $S(t)$ needs to be updated only when an event that changes its state occurs). The triggering events are: (i) a CAV entering the CZ, (ii) a CAV departing the CZ, (iii) a CAV completing a lane change at a floating MP, and (iv) a CAV overtaking event (a lane change event at a fixed MP). This last event may occur when a CAV merges into another lane at an MP through which it leaves the CZ. In particular, consider three CAVs i, j, and k such that $k > j > i$, and CAV j merges into the same lane as i and k. Then, CAV k looks at the first row above it where there is a CAV with the same lane; that's now CAV j. However, i is physically ahead of k. Thus, we need to re-order the queue according to the incremental position order, so that a CAV following i (CAV k) can properly identify its physically preceding CAV. For example, consider $i = 7$, $j = 8$, and $k = 9$ in Fig. 9.4, and suppose that CAV 7 turns right, CAV 8 turns right, and CAV 9 goes straight. Their order in $S(t)$ is 7, 8, 9. CAV 8 can overtake CAV 7, and its current lane will become l_6 when it passes all MPs. Since CAV 7 and CAV 9 also are in l_6, CAV 9 will mistake CAV 8 as its new preceding CAV after the current lane of CAV 8 is updated. However, in reality CAV 7 is still the preceding CAV of CAV 9, hence CAV 9 may accidentally collide with CAV 7. To avoid this problem, we need to re-order the queue according to the position information when this event occurs, i.e., making CAV 8 have higher priority than CAV 7 in the queue. Alternatively, this problem may be resolved by simply neglecting CAVs that have passed all MPs when searching for the correct identity of the CAV that precedes i.

9.3.2.3 Unconstrained Optimal Control

Once a newly arriving CAV $i \in S(t)$ has determined all the lateral safety constraints (9.48) it has to satisfy, it can solve problem (9.51) subject to these constraints along with the rear-end safety constraint (9.47) and the state limitations (9.49). Obtaining a solution to this constrained optimal control problem is computationally intensive. Therefore, proceeding as in the case of the roundabout traffic control problem, we begin by obtaining an analytical solution of problem (9.51) when all state and safety constraints are inactive. This solution is of the same form as (9.35)–(9.37) for the unconstrained optimal control, speed, and position trajectories where a_i, b_i, c_i and d_i are integration constants that can be solved along with t_i^m by the following five algebraic equations (details given in [102]):

$$\frac{1}{2}a_i \cdot (t_i^0)^2 + b_i t_i^0 + c_i = v_i^0,$$

$$\frac{1}{6}a_i \cdot (t_i^0)^3 + \frac{1}{2}b_i \cdot (t_i^0)^2 + c_i t_i^0 + d_i = 0,$$

$$\frac{1}{6}a_i \cdot (t_i^m)^3 + \frac{1}{2}b_i \cdot (t_i^m)^2 + c_i t_i^m + d_i = L_k, \qquad (9.55)$$

$$a_i t_i^m + b_i = 0,$$

$$\beta - \frac{1}{2}b_i^2 + a_i c_i = 0,$$

where the third equation is the terminal condition for the total distance traveled on a lane. This solution is computationally very efficient to obtain and will be used as a reference CAV trajectory in the OCBF approach, similar to the merging and roundabout traffic control problems considered in prior sections.

9.3.2.4 Joint Optimal Control and Control Barrier Function (OCBF) Method

As in the merging and roundabout traffic control problems addressed through the OCBF approach, we proceed in two steps: (i) We solve the *unconstrained* version using only initial and final conditions with a free terminal time in problem (9.51), as described in Sect. 9.3.2.3, and (ii) We optimally track the unconstrained problem solution while using CBFs to account for all constraints and guarantee that they are never violated (as well as Control Lyapunov Functions (CLFs) to better track the unconstrained optimal states).

To accomplish the optimal tracking performed by the OCBF controller, first let $x_i(t) \equiv (x_i(t), v_i(t))$. Referring to the vehicle dynamics (9.45), let $f(x_i(t)) = [x_i(t), 0]^T$ and $g(x_i(t)) = [0, 1]^T$. Each of the constraints in (9.47), (9.48) and (9.49) can be expressed as $b_k(x_i(t)) \geq 0$, $k \in \{1, \cdots, n\}$ where n is the number of constraints and each $b_k(x_i(t))$ is a CBF. For example, $b_1(x_i(t)) = z_{i,i_p}(t) - \varphi v_i(t) - \delta$ for the rear-end safety constraint (9.47). As usual, each of the continuously differentiable *state* constraints $b_k(x_i(t)) \geq 0$ is mapped onto another constraint on the *control* input such that the satisfaction of this new constraint implies the satisfaction of the original constraint $b_k(x_i(t)) \geq 0$. The forward invariance property (see Chap. 2) ensures that a control input that satisfies the new constraint is guaranteed to also satisfy the original one. In particular, each of these new constraints takes the form

$$L_f b_k(\mathbf{x}_i(t)) + L_g b_k(\mathbf{x}_i(t))u_i(t) + \gamma(b_k(\mathbf{x}_i(t))) \geq 0, \tag{9.56}$$

where L_f and L_g denote the Lie derivatives of $b_k(\mathbf{x}_i(t))$ along f and g (defined above from the vehicle dynamics) respectively and $\gamma(\cdot)$ denotes a class $\mathcal{K}$ function (typically, linear or quadratic functions). Similarly, a CLF $V(\mathbf{x}_i(t))$, instead of $b_k(\mathbf{x}_i(t))$, can also be used to track (stabilize) the optimal speed trajectory through a CLF constraint of the form

$$L_f V(\mathbf{x}_i(t)) + L_g V(\mathbf{x}_i(t))u_i(t) + \epsilon V(\mathbf{x}_i(t)) \leq e_i(t), \tag{9.57}$$

where $\epsilon > 0$ and $e_i(t)$ is a relaxation variable that makes this constraint soft. As is usually the case, we select $V(\mathbf{x}_i(t)) = (v_i(t) - v_{ref}(t))^2$ where $v_{ref}(t)$ is the reference speed to be tracked (specified below). Therefore, the OCBF controller solves the following problem:

$$\min_{u_i(t),e_i(t)} J_i(u_i(t), e_i(t)) = \int_{t_i^0}^{t_i^m} \left(\beta e_i^2(t) + \frac{1}{2}(u_i(t) - u_{ref}(t))^2 \right) dt, \tag{9.58}$$

subject to the vehicle dynamics (9.45), the CBF constraints (9.56) and the CLF constraint (9.57). The obvious selection for speed and acceleration reference signals is $v_{ref}(t) = v_i^*(t)$, $u_{ref}(t) = u_i^*(t)$ with $v_i^*(t)$, $u_i^*(t)$ given by (9.36) and (9.35) respectively. In [86], $u_{ref}(t) = \frac{x_i^*(t)}{x_i(t)} u_i^*(t)$ is used to provide the benefit of feedback obtained by observing the actual CAV trajectory $x_i(t)$ and automatically reducing the tracking position error; we use only $u_{ref}(t) = u_i^*(t)$ in the sequel for simplicity.

The CBF conversions from the original constraints to the form (9.56) are straightforward. For example, using a linear function $\gamma(\cdot)$ in (9.56), we can directly map **Constraint 1** onto the following constraint in terms of control inputs:

$$\underbrace{v_{i_p}(t) - v_i(t)}_{L_f b(\mathbf{x}_i(t))} + \underbrace{-\varphi}_{L_g b(\mathbf{x}_i(t))} u_i(t) + z_{i,i_p}(t) - \varphi v_i(t) - \delta \geq 0. \tag{9.59}$$

However, there are some points that deserve some further clarification as follows.

Constraint 2 (*Lateral safety constraint*): The lateral safety constraints in (9.48) are specified only at time instants t_i^k. However, to use CBFs as in (9.56), they have to be converted to *continuously differentiable forms*. Thus, we use the same technique as in Sect. 9.2.2.6 (see also [73]) to convert (9.48) into:

$$z_{i,j}(t) \geq \Phi(x_i(t))v_i(t) + \delta, \quad i \in S(t), \ t \in [t_i^0, t_i^k], \tag{9.60}$$

where $j \in \Omega_i$ is determined through the lateral safety constraint determination strategy (**Algorithm** 2). Recall that CAV j depends on some MP M_k and we may have several $j \in \Omega_i$ since CAV i may conflict with several CAVs j at different MPs. The selection of function $\Phi : \mathbb{R} \to \mathbb{R}$ is flexible as long as it is a strictly increasing function that satisfies $\Phi(x_i(t_i^0)) = 0$ and $\Phi(x_i(t_i^k)) = \varphi$ where t_i^k is the arrival time at MP M_k corresponding to the constraint and $x_i(t_i^k)$ is the location of MP M_k. Thus, we see that at $t = t_i^k$ all constraints in (9.60) match the safe-merging constraints (9.5), and that at $t = t_i^0$ we have $z_{i,i_p}(t_i^0) = \delta$. Since the selection of $\Phi(\cdot)$ is flexible, for simplicity, we define it to have the linear form $\Phi(x_i(t)) = \frac{\varphi}{x_i(t_i^k)} x_i(t)$ which we can immediately see satisfies the properties above.

Improving the feasibility of Constraints 1 and 2: In order to ensure that a feasible solution always exists for these constraints, we need to take the braking distance into consideration. CAV i should stop within a minimal safe distance when its speed $v_i(t)$ approaches the speed $v_j(t)$ for any j such that CAV j is the preceding vehicle of CAV i or any vehicles that may laterally collide with CAV i. Thus, we use the following stricter constraint when $v_i(t) \geq v_j(t)$:

$$z_{i,j}(t) \geq \frac{\varphi\left(x_i(t) + \frac{1}{2}\frac{v_i^2(t) - v_j^2(t)}{u_{\min}}\right)v_j(t)}{L} + \frac{1}{2}\frac{(v_j(t) - v_i(t))^2}{u_{\min}} + \delta, \tag{9.61}$$

A detailed analysis of this constraint is given in [73].

With all constraints converted to CBF constraints in problem (9.51), the solution is obtained, as in the merging and roundabout traffic control problems, through the usual method first presented at the end of Chap. 2: we discretize the time interval $[t_i^0, t_i^m]$ with time steps of length Δ and solve (9.41) over $[t_i^0 + k\Delta, t_i^0 + (k+1)\Delta]$, $k = 0, 1, \ldots$, with $u_i(t), e_i(t)$ as decision variables held constant over each such interval. Consequently, each such problem is a QP since we have a quadratic cost and a number of linear constraints on the decision variables at the beginning of each interval. The solution of each such problem gives an optimal control $u_i^*(t_i^0 + k\Delta)$, $k = 0, 1, \ldots$, allowing us to update (9.45) in the kth time interval. This process is repeated until CAV i leaves the CZ.

9.3.2.5 The Impact of Noise and Complicated Vehicle Dynamics

Aside from the potentially long computation time, other limitations of the original optimal control problem (9.33) include: (i) It only plans the optimal trajectory once. However, the trajectory may violate safety constraints due to noise in the vehicle dynamics and control accuracy; (ii) The analytical solution of (9.33) is limited to simple vehicle dynamics as in (9.45) and becomes difficult to obtain when more complicated vehicle dynamics are considered to better match realistic operating conditions. For instance, in practice, we usually need to control the input driving force of an engine instead of directly controlling acceleration. Compared to solving (9.33), the OCBF approach can effectively deal with the above problems with only slight modifications as described next.

First, due to the presence of noise, constraints may be temporarily violated, which prevents the CBF method from satisfying the forward invariance property. Thus, when a constraint is violated at time t_1, i.e., $b_k(x_i(t_1)) < 0$, we add a threshold to the original constraint as follows:

$$L_f b_k(x_i(t)) + L_g b_k(x_i(t))u_i(t) \geq c_k(t), \tag{9.62}$$

where $c_k(t) \geq 0$ is a large enough value so that $b_k(x_i(t))$ is strictly increasing even if the system is under the worst possible noise case. Since it is hard to directly determine the value of $c_k(t)$, we add it to the objective function and have

$$\min_{u_i(t),e_i(t),c_k(t)} \int_{t_i^0}^{t_i^m} \left(\beta e_i^2(t) + \frac{1}{2}(u_i(t) - u_{ref}(t))^2 - \eta c_k(t) \right) dt, \tag{9.63}$$

where η is a weight parameter. If there are multiple constraints that are violated at one time, we rewrite them all as (9.62) and add all thresholds into the optimization objective. Starting from t_1, we use the constraint (9.62) and objective function (9.63) to replace the original CBF constraint and objective function; then, $b_k(x_i(t))$ will be positive again in finite time since it is increasing. When $b_k(x_i(t))$ becomes positive again, we revert to the original CBF constraint.

Next, considering vehicle dynamics, there are numerous models which achieve greater accuracy than the simple model (9.45) depending on the situation of interest. As an example, we consider the following frequently used nonlinear model:

$$\begin{bmatrix} \dot{x}_i(t) \\ \dot{v}_i(t) \end{bmatrix} = \begin{bmatrix} v_i(t) \\ -\frac{1}{m_i} F_r(v_i(t)) \end{bmatrix} + \begin{bmatrix} 0 \\ \frac{1}{m_i} \end{bmatrix} u_i(t), \tag{9.64}$$

where m_i denotes the vehicle mass and $F_r(v_i(t))$ is the resistance force that is normally expressed as

$$F_r(v_i(t)) = \alpha_0 sgn(v_i(t)) + \alpha_1 v_i(t) + \alpha_2 v_i^2(t), \tag{9.65}$$

where $\alpha_0 > 0$, $\alpha_1 > 0$, and $\alpha_2 > 0$ are parameters determined empirically, and $sgn(\cdot)$ is the signum function. It is clear that due to the nonlinearity in these vehicle dynamics, it is unrealistic to expect an analytical solution for it. However, in the OCBF method, we only need to derive the Lie derivative along these new dynamics and solve the corresponding QP based on these new CBF constraints. For instance, it is easy to see that for these new dynamics, the CBF constraint (9.59) becomes

$$\underbrace{v_{i_p}(t) - v_i(t) + \frac{\varphi F_r(v_i(t))}{m_i}}_{L_f b(x_i(t))} + \underbrace{-\frac{\varphi}{m_i}}_{L_g b(x_i(t))} u_i(t)$$
$$+ z_{i,i_p}(t) - \varphi v_i(t) - \delta \geq 0. \tag{9.66}$$

Thus, the OCBF method can be easily extended to more complicated vehicle dynamics dictated by any application of interest.

9.3.3 Simulation Results for Traffic Control in Intersections

In this section, we demonstrate the effectiveness of the OCBF method in the intersection traffic control problem compared to a state-of-the-art method presented in [103] where CAVs calculate the fastest arrival time to the merging zone (where CAV conflict occurs) first when they enter the CZ and then derive an energy-time-optimal trajectory. This is based on the same objective function (9.51) used in this section. The main differences are: (i) the model in [103] considers the merging (conflict) zone as a whole and imposes the conservative requirement that any two vehicles that have potential conflict cannot be in the merging zone at the same time; (ii) when it plans an energy-time-optimal trajectory for a new incoming vehicle, it takes all safety constraints into account, which makes it time-consuming; (iii) the rear-end safety constraints used in [103] only depend on distance, i.e., $\varphi = 0$ and $\delta > 0$ in (9.47). Thus, in order to carry out a fair comparison with this method, we adopt the same form of rear-end safety constraints, that is,

$$z_{i,i_p}(t) \geq \delta, \quad \forall t \in [t_i^0, t_i^m]. \tag{9.67}$$

A complication caused by this choice is that after using the standard CBF method (simply substituting $\varphi = 0$ into (9.59)), the control input should satisfy

$$\underbrace{v_{i_p}(t) - v_i(t)}_{L_f b(\boldsymbol{x}_i(t))} + \underbrace{0}_{L_g b(\boldsymbol{x}_i(t))} u_i(t) + z_{i,i_p}(t) - \delta \geq 0, \tag{9.68}$$

which violates the condition $L_g b(\boldsymbol{x}_i(t)) \neq 0$. This is because we cannot obtain a relationship involving the control input $u_i(t)$ from the first-order derivative of the constraint (9.67), a problem we dealt with in Chap. 3 by using HOCBFs. In this case, we use a HOCBF of relative degree 2 for system (9.45). In particular, letting $b_k(\boldsymbol{x}_i(t)) = z_{i,i_p}(t) - \delta$ and considering all class $\mathcal{K}$ functions to be linear functions, we define

$$\begin{aligned} \psi_1(\boldsymbol{x}_i(t)) &= \dot{b}(\boldsymbol{x}_i(t)) + pb(\boldsymbol{x}_i(t)), \\ \psi_2(\boldsymbol{x}_i(t)) &= \dot{\psi}_1(\boldsymbol{x}_i(t)) + p\psi_1(\boldsymbol{x}_i(t)). \end{aligned} \tag{9.69}$$

where p is a tunable penalty coefficient. Combining the vehicle dynamics (9.45) with (9.69), any control input should satisfy

$$\underbrace{0}_{L_f^2 b(\boldsymbol{x}_i(t))} + \underbrace{-1}_{L_g L_f b(\boldsymbol{x}_i(t))} u_i(t) + 2p\dot{b}(\boldsymbol{x}_i(t)) + p^2 b(\boldsymbol{x}_i(t)) \geq 0. \tag{9.70}$$

Thus, in the following simulation experiments, we set φ in (9.47) for **Constraint 1** and in (9.48) for **Constraint 2** to be $\varphi = 0$ and $\varphi = 1.8s$, respectively.

Our simulation experiments are organized as follows. First, in Sect. 9.3.3.1 we consider an intersection with a single lane in each direction which only allows a CAV to go straight. Our purpose here is to show that using MPs instead of an entire arbitrarily defined merging zone can effectively reduce the conservatism of the latter. Then, in Sect. 9.3.3.2, we allow turns so as to analyze the influence of different behaviors (going straight, turning left, and turning right) on the performance of the methods compared. Next, Sect. 9.3.3.3 is intended to validate the effectiveness of our OCBF method for intersections with two lanes and include possible lane-changing behaviors. Section 9.3.3.4 shows simulations that quantify how our proposed method outperforms the signal-based method in all metrics. In Sect. 9.3.3.5, we extend our method to combine it with the Dynamic Resequencing (DR) method and show that the performance of the combined OCBF+DR method is better than the OCBF+FIFO method for asymmetrical intersections. Finally, Sect. 9.3.3.6 demonstrates how the OCBF method can effectively deal with complicated vehicle dynamics and noise.

The baseline for our simulation results uses SUMO, a microscopic traffic simulation software package (in contrast to using the Vissim simulator as in earlier sections). In SUMO simulations, we set most models and their parameters at their default values. For example, the car-following model used in SUMO is the default Krauss model, and its parameters are also set at their default values. Interested readers can refer to [34] for more details. Then, we use our OCBF controller and the controller in [103] to control CAVs for intersection scenarios with the same vehicle arrival patterns as SUMO. The parameter settings (see Fig. 9.4) are as follows: $L_1 = 300\,\text{m}$, $L_2 = 50\,\text{m}$, $L_3 = 200\,\text{m}$, $l = 0.9378\,\text{m}$, $w = 3.5\,\text{m}$, $r = 4\,\text{m}$, $\delta = 10\,\text{m}$, $v_{\max} = 15\,\text{m/s}$, $v_{\min} = 0\,\text{m/s}$, $u_{\max} = 3\,\text{m/s}^2$, and $u_{\min} = -3\,\text{m/s}^2$.

The energy model we used in the objective function (9.51) is an approximate one. The $\frac{1}{2}u^2$ metric treats acceleration and deceleration the same and does not account for speed as contributing to energy consumption. This metric is viewed as a simple surrogate function for energy or simply as a measure of how much the solution deviates from the ideal constant-speed trajectory. In contrast, the following energy model [24] captures fuel consumption in detail and provides another measure of performance:

$$
\begin{aligned}
F_i &= \int_0^{a_i} f_{V,i}(t)\,dt, \\
f_{V,i}(t) &= f_{cruise,i}(t) + f_{accel,i}(t), \\
f_{cruise,i}(t) &= b_0 + b_1 v_i(t) + b_2 v_i^2(t) + b_3 v_i^3(t), \\
f_{accel,i}(t) &= u(t)(c_0 + c_1 v_i(t) + c_2 v_i^2(t)),
\end{aligned}
\tag{9.71}
$$

where $f_{cruise,i}(t)$ denotes the fuel consumed per second when CAV i drives at a steady velocity $v_i(t)$, and $f_{accel,i}(t)$ is the additional fuel consumed due to the presence of positive acceleration. If $u(t) \leq 0$, then $f_{accel,i}(t)$ will be 0 since, in this case, the engine is rotated by the kinetic energy of the CAV. b_0, b_1, b_2, b_3, c_0, c_1, and c_2 are seven model parameters;

Table 9.9 Comparison results for a single-lane intersection disallowing turns

β	Methods	Energy	Travel time (s)	Fuel (mL)	Ave. obj.[1]
0.1	SUMO	23.1788	28.3810	30.3597	26.0169
	OC	0.1498	28.3884	14.6266	2.9886
	OCBF	0.9501	25.0863	18.3088	3.4587
0.5	SUMO	23.1788	28.3810	30.3597	37.3693
	OC	0.6515	26.0315	17.1585	13.6673
	OCBF	2.1708	22.6623	18.9396	13.5020
1	SUMO	23.1788	28.3810	30.3597	51.5598
	OC	0.8782	25.6961	17.1555	26.5743
	OCBF	2.9106	22.2617	18.9589	25.1723
2	SUMO	23.1788	28.3810	30.3597	79.9408
	OC	1.1869	25.4834	17.1658	52.1537
	OCBF	3.9157	21.9139	18.9852	47.7435

[1] Ave. obj. $= \beta \times$ Travel time $+$ Energy

here we use the same parameters as in [24], which are obtained through curve-fitting for data from a typical vehicle.

9.3.3.1 Merging Points Versus a Merging Zone

We start by allowing CAVs to only go straight in order to investigate the relative performance of the OCBF controller based on Merging Points (MPs) and the method used in [103] (OC controller) that uses a Merging Zone (MZ). We set the arrival rates at all lanes to be the same, i.e., 270 veh/h/lane. The comparison results are shown in Table 9.9.

It is clear that both controllers significantly outperform the results obtained from the SUMO car-following controller. The OC controller is energy-optimal since it has considered all safety constraints for each CAV upon its arrival at the CZ. Subsequently, CAVs strictly follow the planned trajectory assuming the absence of noise. However, the OCBF controller only uses an *unconstrained* reference trajectory and employs CBFs to account for the fact that this reference trajectory may violate the safety constraints: for each CAV, this controller continuously updates its control inputs according to the latest states of other CAVs. As a result, its energy consumption is larger than that of the OC controller, although still small and much lower than the one evaluated under the SUMO car-following controller.

In terms of travel time, we find that the travel time of the OCBF controller is better than that of the OC controller. This is because the safety requirements in the OC controller are too strict because CAV i must wait until a CAV $j \neq i$ that conflicts with it leaves the MZ. Instead, the OCBF controller using MPs allows us to relax a merging constraint and still ensure safety by requiring that only one CAV can arrive at the same MP φ a short time after

Table 9.10 Influence of turns on different controllers

Groups	Methods	Energy	Travel time (s)	Fuel (mL)	Ave. obj.
1	SUMO	23.7011	28.2503	28.3357	51.9514
	OC	1.2957	22.4667	18.8433	23.7624
	OCBF	2.6193	21.8514	18.9208	24.4707
2	SUMO	26.5612	31.3822	29.0524	57.9434
	OC	1.0666	24.2179	18.0072	25.2845
	OCBF	2.4537	21.8707	18.8704	24.3244
3	SUMO	19.9066	24.2937	25.6778	44.2003
	OC	1.3775	22.0706	18.8803	23.4481
	OCBF	2.3623	21.4874	18.8306	23.8497
4	SUMO	21.7450	26.6148	29.4484	48.3598
	OC	1.2602	22.7417	18.7339	24.0019
	OCBF	2.5884	22.0114	18.9230	24.5998

the other vehicle leaves. Since the OCBF method reduces conservatism, it shows significant improvement in travel time when compared with the OC controller in [103].

In addition, we can adjust the parameter β to emphasize the relative importance of one objective (energy or time) relative to the other. If we are more concerned about energy consumption, we can use a smaller value of β; otherwise, a larger β value emphasizes travel time reduction. Thus, when β is relatively large, the average objective under the OCBF controller is better than that of the OC controller since it is more efficient with respect to travel time.

Another interesting observation is that even though the relationship between the accurate fuel consumption model and the estimated energy is complicated, we see that a larger estimated energy consumption usually corresponds to larger fuel consumption. Thus, it is reasonable to optimize energy consumption through a simple model, e.g., $\frac{1}{2}u^2$, which also significantly reduces the computational complexity caused by the accurate energy model.

9.3.3.2 The Influence of Turns

We now allow turns at the intersection assuming that the intention (i.e., going straight, turning left, and turning right) of a CAV when it enters the CZ is known. We have conducted four groups of simulations as shown in Table 9.10. In the first group, all CAVs choose their behavior with the same probability, i.e., $\frac{1}{3}$ going straight, $\frac{1}{3}$ turning left, and $\frac{1}{3}$ turning right. In the second group, 80% of CAVs turn left while 10% CAVs go straight and 10% CAVs turn right. In the third group, 80% CAVs turn right while 10% CAVs go straight and 10% CAVs turn left. In the fourth group, 80% CAVs go straight while 10% CAVs turn left and 10% CAVs turn right. We set $\beta = 1$ in all results shown in Table 9.10.

Table 9.11 Comparison results for a two-lane intersection

β	Methods	Energy	Travel time (s)	Fuel (mL)	Ave. obj.
0.1	SUMO	24.0124	29.5955	30.5588	26.9720
	OC	0.1514	28.5711	14.6685	3.0085
	OCBF	1.0350	25.0804	18.5086	3.5430
0.5	SUMO	24.0124	29.5955	30.5588	38.8102
	OC	0.6722	26.1284	17.2866	13.7364
	OCBF	2.2244	22.6351	19.0753	13.5420
1	SUMO	24.0124	29.5955	30.5588	53.6079
	OC	0.9063	25.8000	17.2933	26.7063
	OCBF	2.9955	22.2347	19.1126	25.2302
2	SUMO	24.0124	29.5955	30.5588	83.2034
	OC	1.2294	25.5773	17.3042	52.3840
	OCBF	4.2353	22.1167	19.1500	48.4687

First, we can draw the same conclusion as in Table 9.9 that the OC controller is energy-optimal and the OCBF controller achieves the lowest travel time since it reduces conservatism. Next, we also observe that when we increase the ratio of left-turning vehicles, the average travel times under all controllers increase; when we increase the ratio of right-turning vehicles, the average travel times all decrease. This demonstrates that the left-turning behavior usually has the largest impact on traffic coordination since left-turning CAVs cross the conflict zone diagonally and are more likely to conflict with other CAVs. In addition, it is worth noting that going straight produces the largest travel time since this involves the largest number of MPs. However, when we use the OCBF controller, the travel times under all situations are similar, which shows that this controller can utilize the space resources of the conflict zone and handle the influence of turns more effectively.

9.3.3.3 Comparison Results for More Complicated Intersections

In this set of simulation experiments, we consider more complicated intersections with two lanes in each direction as shown in Fig. 9.4. The left lane in each direction only allows going straight and turning left, while the right lane only allows going straight and turning right. We set the arrival rate at all lanes to be the same, i.e., $180veh/h/lane$ and disallow lane-changing. Each new incoming CAV chooses its behavior from the allowable behaviors with the same probability, e.g., the CAV arriving at the entry of the left lane can go straight or turn left with probability 0.5. The comparison results are shown in Table 9.11.

The results here are similar to those in the single-lane intersections. Although the number of MPs increases with the number of lanes, the OCBF method can still effectively ensure safety and reduce travel time. It is worth noting that the values of safety time headway for the OC controller are difficult to determine. The OC controller requires that no CAV can enter the MZ until the conflict CAV leaves it. However, the time spent for passing through the MZ differs from vehicle to vehicle. If we choose a larger value, then this significantly increases travel time and amplifies conservatism. In contrast, if we choose a smaller value, the potential of collision increases. Therefore, the MP-based method is significantly better since it not only ensures safety but also reduces conservatism.

In what follows, we briefly discuss the computation time involved in the OCBF method. The method is driven by solving a QP problem using the latest information from all CAVs over each control period. The average computation time for solving such a QP is 3.5 ms (Intel(R) Core(TM) i7-6700 CPU), which is suitable for real-time implementation. In contrast, the OC-based controller plans a trajectory for each vehicle when it enters the control zone, and the average computation time for such a plan is about 1 s (Intel(R) Core(TM) i7-6700 CPU). Furthermore, when multiple constraints become active, the computation time ranges from 3 to 30 s (Intel(R) Core(TM) i7-8700 CPU). Thus, the OCBF method is more suitable for complicated vehicle dynamics and constraints since its computation time is not affected by these factors.

Next, we consider the impact of lane-changing on the OCBF method. For the same two-lane intersection, we allow lane-changing and CAVs can choose any movement (going straight, turning left and right). Since the left lane only allows going straight and turning left, the right-turning CAV in this lane must change its lane. The situation is similar for the left-turning CAV in the right lane. To make a better comparison with the scenario without lane-changing, we use the same arrival data (including the times all CAVs enter the CZ and initial velocities) as the last experiment and only change the lane that the turning CAV arrives at. For example, the left-turning CAVs must arrive at the left lane in the last experiment, but, in this experiment, the lane they enter can be random. The results are shown in Table 9.12.

We can see that the lane-changing behavior slightly increases all performance measures compared with the results in scenarios disallowing lane-changing. This is expected since a new (floating) MP is added and more control is required to ensure safety. Nevertheless, the changes are minor, fully demonstrating the effectiveness of the OCBF method in handling lane changing. Although we have assumed that lane changing only induces a fixed length, we can extend the OCBF method to more complicated lane-changing trajectories, e.g., trajectories fitted by polynomial functions. Note that in the SUMO simulation, it is assumed that a vehicle can jump directly from one lane to another. However, the OCBF method still outperforms it in all metrics, further supporting the advantages of the OCBF controller.

Next, we explore the effect of asymmetrical arrival rates through two scenarios, in order to confirm that the OCBF method is effective even when traffic flows are heavy. In the first scenario, we set the arrival rates in Lanes 1, 2 to be three times as large as Lanes 3–8; while in the second scenario, the arrival rates in Lanes 1, 2, 5, 6 are three times as large as the remaining lanes. The comparison results are shown in Table 9.13.

We can see in the SUMO simulation that traffic flows in lanes with high arrival rates are highly congested with CAVs forming long queues in these lanes. All metrics obtained from

Table 9.12 Influence of lane-changing behavior on OCBF method

β	Methods	Energy	Travel time (s)	Fuel (mL)	Ave. obj.
0.1	SUMO with LC	23.9988	30.0337	30.0000	27.0022
	OCBF w/o LC	1.0350	25.0804	18.5086	3.5430
	OCBF with LC	1.0738	25.1200	18.5474	3.5858
0.5	SUMO with LC	23.9988	30.0337	30.0000	39.0157
	OCBF w/o LC	2.2244	22.6351	19.0753	13.5420
	OCBF with LC	2.2584	22.6689	19.1148	13.5929
1	SUMO with LC	23.9988	30.0337	30.0000	54.0325
	OCBF w/o LC	2.9955	22.2347	19.1126	25.2302
	OCBF with LC	3.0282	22.2684	19.1575	25.2966
2	SUMO with LC	23.9988	30.0337	30.0000	84.0662
	OCBF w/o LC	4.2353	22.1167	19.1500	48.4687
	OCBF with LC	4.2887	22.1536	19.2457	48.5959

Table 9.13 Influence of asymmetrical and heavy traffic flows

Scenario	Methods	Energy	Travel time (s)	Fuel (mL)	Ave. obj.
1	SUMO	43.1656	68.6488	40.4731	111.8144
	OCBF	2.6312	22.1812	19.1493	24.8124
2	SUMO	44.4815	96.0517	44.6265	140.5332
	OCBF	3.0999	22.5594	19.5380	25.6593

Scenario 1: arrival rates at l_1 and l_2 are 540 veh/h/lane; while at l_3 to l_8 they are 180 veh/h/lane.
Scenario 2: arrival rates at l_1, l_2, l_5, and l_6 are 540 veh/h/lane; while at l_3, l_4, l_7, and l_8 they are 180 veh/h/lane

SUMO significantly increase compared with the results obtained from medium traffic shown in Table 9.11. However, since the coordination performance under the OCBF controller is much better than SUMO, all metrics remain at low levels, indicating the effectiveness of the OCBF approach in congested situations.

9.3.3.4 Comparison Results with an Actuated Signal-based Traffic Controller

To further demonstrate that the OCBF method can achieve more efficient use of road resources, we carry out a comparison with a baseline scenario under the control of adaptive four-phase traffic lights using SUMO. In this baseline scenario, an intersection has two lanes in each direction as shown in Fig. 9.4. It is also worth noting that SUMO employs a gap-based actuated traffic control method whose main idea is to prolong a traffic phase when detecting a continuous stream of traffic. In this simulation, all model parameters used are set at their default values. Then, to investigate the influence of different traffic volumes, we vary the arrival rates to generate different traffic demands. The comparison results are shown in Fig. 9.5.

The simulation results show that when the traffic volume is low, the actuated traffic control method tends to shorten each phase's duration to reduce the waiting time of vehicles before the stop line. However, we can still observe the undesirable phenomenon that a vehicle is waiting for a green light in order to pass, while the intersection is empty. In contrast, when the traffic volume is high, the actuated traffic control method makes each phase's duration as long as possible to avoid interrupting the traffic flow in that direction. However, vehicles need to wait for an excessive amount of time before the stop line in this situation. Compared with the signal-based method, our optimized signal-free-based method is significantly better in all metrics. Although its performance deteriorates when traffic congestion builds up, the changes in metric values are relatively small. Based on the SUMO simulations, the maximum traffic volume the intersection can support under the signal-based method is approximately 3000 veh/h (or 375 veh/h/lane). When we continue to increase the arrival rates, almost all roads are jammed so that new vehicles cannot enter the control zone. Thus, the throughput

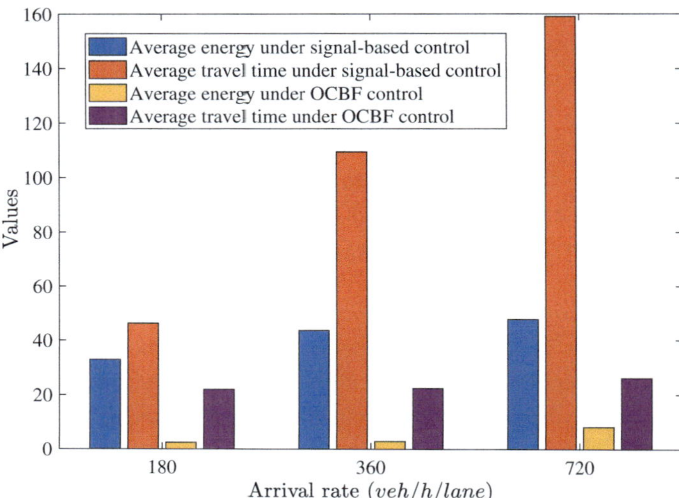

Fig. 9.5 Comparison results between the OCBF method and the actuated signal-based method

Table 9.14 The effect of the DR method on the OCBF controller

Scenario	Methods	Energy	Travel time (s)	Fuel (mL)	Ave. obj.
1	OCBF+FIFO	5.1261	21.6973	19.1518	113.6126
	OCBF+DR	5.1439	21.6404	18.9364	113.3459
2	OCBF+FIFO	5.9218	20.8093	18.6319	109.9683
	OCBF+DR	6.1080	20.6102	18.9077	109.1590
3	OCBF+FIFO	8.9344	19.3548	17.7396	105.7084
	OCBF+DR	7.0501	17.3253	17.2104	93.6766

Scenario 1 is a symmetrical intersection with all lanes are 300 m
Scenario 2 is an asymmetrical intersection with lane 3 and 4 are 200 m while lane 1, 2, 5, 6, 7, and 8 are 300 m
Scenario 3 is an asymmetrical intersection with lane 3 and 4 are 200 m, lane 5 and 6 are 100 m, while lane 1, 2, 7, and 8 are 300 m

values when arrival rates are 360 and 720 veh/h/lane do not differ much. Conversely, traffic flows in all directions are still smooth under the control of the proposed method when the arrival rate is 360 veh/h/lane, hence effectively increasing the intersection throughput.

9.3.3.5 The Inclusion of Dynamic Resequencing

Thus far in all simulation experiments, the FIFO-based queue was used to determine the passing priority when potential conflicts occur. We now combine the OCBF controller with a typical resequencing policy, in particular the Dynamic Resequencing (DR) method [102]. When a new CAV enters the CZ, the DR policy inserts it into the optimal position of the original crossing sequence. Note that when combining the OC method with a resequencing policy, an update of the arrival times and trajectories of CAVs is required whenever we adjust the original crossing sequence. However, in the OCBF method, CAV i only needs to update the indices of the CAVs with which it conflicts according to the new DR-based queue and follow the original unconstrained optimal trajectory without replanning. In the following simulations, we set $\beta = 5$ and vary the length of some lanes to generate different scenarios. The comparison results are shown in Table 9.14.

The DR method helps decrease travel times and achieves a better average objective value at the expense of energy consumption, since CAVs need to take more acceleration/deceleration actions to adjust their crossing order. The benefits of the DR method relative to the FIFO policy are more evident in asymmetrical intersections. This is because the FIFO rule may require a CAV that enters the CZ later but is much closer to the intersection to yield to a CAV that is further away from the intersection. For example, in the above Scenario 3, a CAV enters the CZ from lane 5 that is 100 m away from the intersection. It is unreasonable to force it to yield to a CAV entering earlier but located 250 m away from the intersection. Our resequencing method can effectively avoid such situations by adjusting crossing sequences in an event-driven way. Note that the OCBF+DR method outperforms the OCBF+FIFO

Table 9.15 Influence of nonlinear vehicle dynamics on the OCBF method

β	Energy	Travel time (s)	Fuel (mL)	Ave. obj.
0.1	0.4751	24.5680	16.6496	2.9319
0.5	1.6822	22.3889	18.2822	12.8767
1	2.4702	21.9973	18.4682	24.4675
2	3.4667	21.7871	18.5832	47.0409

method in all metrics in Scenario 3, since nearly all CAVs arriving at lanes 5 and 6 need to decelerate and even stop due to the FIFO rule, indicating that the DR method is more effective when an intersection geometric configuration is asymmetrical. This finding is consistent with the conclusion given in [93] which provides detailed and comprehensive comparisons for different resequencing methods.

9.3.3.6 The Influence of Nonlinear Vehicle Dynamics and Noise

We now consider the nonlinear vehicle dynamics in (9.64) and reformulate all CBF constraints according to the new dynamics. For the symmetrical intersection with two lanes in each direction, we vary β from 0.1 to 2 and use the OCBF+FIFO method to coordinate the movements of CAVs. The results are shown in Table 9.15.

It is clear that the results conform to the results for the double integrator vehicle dynamics (9.45). When β increases, we are more concerned about the travel time, thus travel time decreases while the energy and fuel consumption rise. Note that although the nonlinear vehicle dynamics are more complicated than the double integrator vehicle dynamics, the only necessary modification is to derive the CBF constraints based on the new dynamics. The computation times for these two different dynamics are nearly the same.

Next, we have considered both noise and nonlinear dynamics. Due to the measurement errors of sensors and imperfect actuators, there exists random noise in position, velocity, and control inputs. To analyze the influence of noise to the OCBF method, we consider uniformly distributed noise processes ($w_{i,p}(t)$ for the position of CAV i, $w_{i,v}(t)$ for the velocity, and $w_{i,u}(t)$ for the control inputs) for this simulation. We set $\beta = 0.1$ and use the OCBF+FIFO method for all experiments. The results are shown in Table 9.16.

The results show that the measurement errors of positions and velocities significantly increase energy consumption. This is because noise causes CAVs to misjudge their states, thus necessitating additional control actions. For example, suppose CAV i is following CAV j and their actual distance is $10m$ at some time point, but, due to noise, CAV i may misjudge this distance to be $8m$, therefore decelerating to enlarge their relative spacing. Then, at the next time point, it may accelerate to keep a desired inter-vehicle space. These frequent acceleration/deceleration maneuvers cause a considerable waste of energy. As uncertainty increases, more control effort is needed to ensure the safety of CAVs when the number

Table 9.16 Influence of noise on the OCBF method

Noise	Energy	Travel time	Fuel	Ave. obj.
No noise	0.4751	24.5680	16.6496	2.9319
$w_{i,u}(t) \in [-0.5, 0.5]$	0.5777	24.6067	16.8714	3.0384
$w_{i,p}(t) \in [-1, 1]$ $w_{i,v}(t) \in [-1, 1]$	5.4662	24.8587	22.3935	7.9521
$w_{i,p}(t) \in [-1, 1]$ $w_{i,v}(t) \in [-1, 1]$ $w_{i,u}(t) \in [-0.5, 0.5]$	5.5723	24.8458	22.3933	8.0569
$w_{i,p}(t) \in [-2, 2]$ $w_{i,v}(t) \in [-2, 2]$ $w_{i,u}(t) \in [-0.5, 0.5]$	31.3250	24.5667	34.1352	33.7817

of noise sources increases and the noise magnitudes goes up. Note that CAVs may even collide with other CAVs when we continuously increase the magnitude of noise. However, when noise is limited, the OCBF method can effectively handle it and does not add any computational burden.

Applications to Robotics **10**

10.1 Ground Robot Obstacle Avoidance

We begin with the standard unicycle model commonly used to capture the motion of a wheeled mobile robot:

$$\dot{x} = v\cos\theta, \quad \dot{y} = v\sin\theta, \quad \dot{v} = u_2, \quad \dot{\theta} = u_1, \tag{10.1}$$

where (x, y) denotes the two-dimensional location, θ is the heading angle, v denotes the linear speed, and u_1, u_2 are the two control inputs (turning speed and forward acceleration). Note that the dynamics are in the form (2.11), with $\boldsymbol{x} = (x, y, \theta, v)^T$, $\boldsymbol{u} = (u_1, u_2)^T$, $f = (v\cos\theta, v\sin\theta, 0, 0)^T$, and $g = (0, 0; 0, 0; 1, 0; 0, 1)$.

A typical navigation task involves two objectives subject to two constraints as follows:

Objective 1 (*Minimize energy consumption*) The goal here is to minimize the total control effort captured as the energy metric

$$J(\boldsymbol{u}(t)) = \int_{t_0}^{t_f} (u_1^2(t) + u_2^2(t))dt. \tag{10.2}$$

Objective 2 (*Reach destination*) The robot is required to reach a given point $(x_d, y_d) \in \mathbb{R}^2$ over a time interval $[t_1, t_2]$, $t_0 \leq t_1 \leq t_2 \leq t_f$.

Constraint 1 (*Safety*) The robot must avoid a circular obstacle described by

$$(x(t) - x_o)^2 + (y(t) - y_o)^2 \geq r^2,$$

where $(x_o, y_o) \in \mathbb{R}^2$ denotes the location of the obstacle and $r = 7\,\text{m}$ is its size (a little larger than the actual size, which is $6\,\text{m}$).

Constraint 2 (*Robot limitations*) The robot's motion is subject to the constraints

© The Author(s), under exclusive license to Springer Nature Switzerland AG 2023

W. Xiao et al., *Safe Autonomy with Control Barrier Functions*,

Synthesis Lectures on Computer Science,

https://doi.org/10.1007/978-3-031-27576-0_10

$$v_{min} \leq v(t) \leq v_{max},$$

$$u_{1,min} \leq u_1(t) \leq u_{1,max}, \quad u_{2,min} \leq u_2(t) \leq u_{2,max},$$

where $v_{min} = 0\,\text{m/s}$, $v_{max} = 2\,\text{m/s}$, $u_{1,max} = -u_{1,min} = 0.2\,\text{rad/s}$, and $u_{2,max} = -u_{2,min}$ $= 0.5\,\text{m/s}^2$.

We use HOCBFs to (strictly) impose **Constraint 1** and **Constraint 2**. In addition, we define two CLFs:

$$V_1(\boldsymbol{x}) = \left(\theta - atan\left(\frac{y_d - y}{x_d - x}\right)\right)^2, \quad V_2(\boldsymbol{x}) = (v - v_d)^2, \tag{10.3}$$

with $v_d = 2\,\text{m/s}$ in order to achieve **Objective 2**. As for **Objective 1**, we capture it by formulating an optimization problem aimed at minimizing the cost (10.2).

For **Constraint 1**, we define a HOCBF:

$$b(\boldsymbol{x}) = (x(t) - x_o)^2 + (y(t) - y_o)^2 - r^2, \tag{10.4}$$

whose relative degree is $m = 2$. Since $L_f^2 b(\boldsymbol{x}) = 2v^2$ is guaranteed to be non-negative, the penalty method presented in Chap. 4 always works given proper $\boldsymbol{x}(t_0)$. However, we use the parameterization method described in Chap. 5, since we also wish to minimize the HOCBF value when the HOCBF constraint first becomes active. This approach gives good results in an unknown environment with obstacles, as shown below.

We use two HOCBFs to impose the speed part of **Constraint 2**:

$$b_{max}(\boldsymbol{x}) = v_{max} - v, \quad b_{min}(\boldsymbol{x}) = v - v_{min}, \tag{10.5}$$

which both have relative degree 1, and we get $L_f b_{max}(\boldsymbol{x}) = L_f b_{min}(\boldsymbol{x}) = 0$. If we set $\boldsymbol{u} = \boldsymbol{0}$, the speed will not change, which will not violate the speed limitations if they are initially satisfied. Therefore, these HOCBF constraints do not conflict with the control bound, and we do not need to use either the penalty or parameterization methods.

We now consider a worst case scenario for which we can easily determine penalties p_1, p_2 and powers q_1, q_2 in applying the parameterization method of Chap. 5. Thus, we select a maximum initial speed $v(t_0) = v_{max}$ and an initial position $(x(t_0), y(t_0)) = (5, 25\,\text{m})$. We assume the obstacle center $(x_o, y_o) = (32, 25\,\text{m})$ and the destination $(x_d, y_d) = (45, 25\,\text{m})$ are aligned, and the initial heading $\theta(t_0) = 0$ is also parallel to this line, as seen in Fig. 10.1a. In what follows, we study the feasibility robustness of the solution (i.e., how feasibility is affected by changes in the state and/or environment) following the definitions and method-ologies presented in Chap. 5.

When the destination component y_d is exactly $25\,\text{m}$, the robot stops before the obstacle (if p_1, p_2, q_1, q_2 are feasible for the QP), i.e., it cannot arrive at the destination. We call this stop point an "equilibrium point," as shown in Fig. 10.1a. However, if y_d has a positive offset (arbitrary small), the robot can bypass the obstacle and arrive at the destination following the left trajectories shown in Fig. 10.1a. Otherwise, the robot will produce right trajectories also

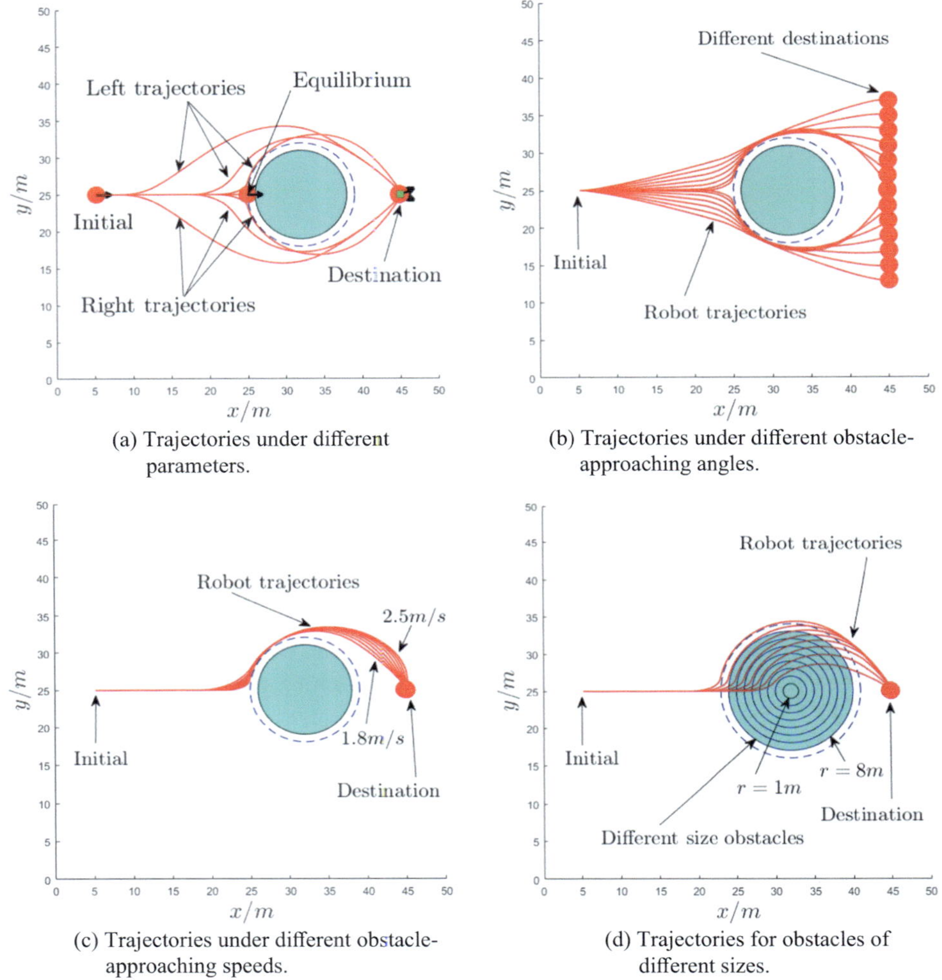

(a) Trajectories under different parameters.

(b) Trajectories under different obstacle-approaching angles.

(c) Trajectories under different obstacle-approaching speeds.

(d) Trajectories for obstacles of different sizes.

Fig. 10.1 Robot obstacle avoidance problem using HOCBFs and the parameterization method

shown in Fig. 10.1a. Therefore, we choose a small offset for y_d (i.e., $y_d = 25.0000001$ m) when trying to find the optimal p_1, p_2, q_1, q_2.

We randomly sample p_1, p_2, q_1, q_2 ($p_1, p_2 \in (0, 3]$, $q_1, q_2 \in [1, 3]$) to get 2000 points, and run simulations for 30s. We use the FGO algorithm in Algorithm 1 shown in Chap. 5 to optimize each sample and get the optimal $(p_1^*, p_2^*, q_1^*, q_2^*) = (0.7535, 0.6664, 1.0046, 1.0267)$ such that the associated QPs are feasible and the HOCBF value is minimized when the HOCBF constraint first becomes active. The HOCBF constraint is active when $b(x) = \mathcal{D}_{min}$ ($\mathcal{D}_{min} = 5.4$ m^2, a distance metric instead of the real distance).

In the rest of this section we study the feasibility robustness of the proposed method, assuming the optimal $(p_1^*, p_2^*, q_1^*, q_2^*)$ given above. Note that the QP feasibility does not depend on the specific initial condition as long as the robot initially has a distance (in terms of $b(x)$) of at least $\mathcal{D}_{min}$ from the obstacle, and the initial state is within the predefined bound (see [74]).

10.1.1 Feasibility Robustness

Feasibility Robustness to the Heading Angle In this case, we only change the value of the destination component y_d. Based on $y_d = 25.0000001$ m and $y_d = 24.9999999$ m, we further offset y_d by $+2$ m for $y_d = 25.0000001$ m (-2 m for $y_d = 24.9999999$ m), and generate 7 destinations for both cases, respectively. These 14 destinations are all feasible, which shows good feasibility robustness of the penalty method to changes in the heading angle when approaching the obstacle, as shown in Fig. 10.1b. The HOCBF values when the HOCBF constraint becomes active are all smaller than $\mathcal{D}_{min}$.

Feasibility Robustness to the Obstacle Approaching Speed Next, we vary the approaching speed to the obstacle between 1.8 and 2.5 m/s (2 m/s was the value for the original problem). All these values all feasible, which shows good feasibility robustness of the penalty method to the change in speed when approaching the obstacle. The HOCBF values when the HOCBF constraint becomes active increase as the approaching speed increases, as shown in Fig. 10.1c.

Feasibility Robustness to the Obstacle Size Here, we only change the obstacle size from the predefined value $r = 7$ m. We consider a range of r between 2 and 9 m. The results show good feasibility robustness to the change of obstacle size as the QPs are always feasible and the robot can safely arrive at its destination, as shown in Fig. 10.1d. The HOCBF values when the HOCBF constraint becomes active do not change under different-size obstacles.

10.1.2 Safe Exploration in an Unknown Environment

Finally, in order to show that the feasibility robustness is independent of the location of the obstacles, we present an application of robot safe exploration in an unknown environment. Suppose the robot is equipped with a sensor ($\frac{2}{3}\pi$ field of view (FOV) and 7 m (greater than the one corresponding to $\mathcal{D}_{min}$) sensing distance with 1 m sensing uncertainty) to detect the obstacles, and there are three unknown obstacles (to the robot) whose center locations are (32, 25 m), (28, 35 m), (30, 40 m) with radius 6, 5, 6 m, respectively. The robot is required to arrive sequentially at points $a := (39, 35$ m$)$, $b := (30, 15$ m$)$, $c := (38, 40$ m$)$, $d := (20, 28$ m$)$. The robot can safely arrive at these four destinations with the penalties and powers $(p_1, p_2, q_1, q_2) = (0.7535, 0.6664, 1.0046, 1.0267)$ as calculated above, which shows good

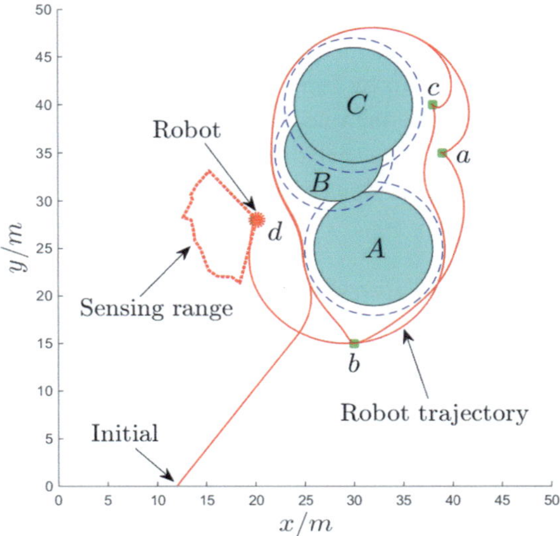

Fig. 10.2 Safe exploration in an unknown environment

feasibility robustness. The robot trajectory is shown in Fig. 10.2. The computation time to solve the QP at each time step is less than $0.01\,$s (Intel(R) Core(TM) i7-8700 CPU @ $3.2\,$GHz$\times$2).

10.2   Rule-Based Autonomous Driving

Autonomous driving is a prominent example of a cyber-physical system which is safety-critical. The proliferation of such systems has given rise to an increasing need for computational tools that can facilitate and enable their verification and control subject to rich and complex specifications. In the case of autonomous driving, there are common objectives for Autonomous Vehicles (AVs) which are captured in optimal control problems, such as minimizing energy consumption and travel time, and constraints on control variables, such as a maximum acceleration. In addition, however, AVs need to follow complex and possibly conflicting traffic laws with different priorities. They should also meet culture-dependent expectations of reasonable driving behavior [17, 23, 45, 47, 50, 56, 65]. For example, an AV has to avoid collisions with other road users (higher priority), maintain longitudinal clearance with respect to a lead vehicle (lower priority), and drive faster than the minimum speed limit (still lower priority). Inspired by [13], we formulate these behavior specifications as a set of rules with a priority structure that captures their importance.

To accommodate the various rules an AV should follow, in this section we formulate an optimal control problem in which the satisfaction of the rules and some vehicle limitations are enforced by CBFs, while CLFs are used to achieve convergence to desired states, similar to the traffic control problems considered in Chap. 9. On the other hand, to minimize the violation of the rules, we formulate iterative rule relaxations according to their priorities. The online implementation of the rule-based optimal control framework can be found in [89]. We begin by formulating the basic rule-based control problem in Sect. 10.2.1, and then introduce the rules and their priority structure in Sect. 10.2.2. The rule-based control framework is presented in Sect. 10.2.3, and several case studies under different driving scenarios are presented in Sect. 10.2.4.

10.2.1 Problem Formulation

For a vehicle with dynamics given by (2.11) and starting at a given state $x(0) = x_0$, consider an optimal control problem in the form:

$$\min_{u(t)} \int_0^T J(||u(t)||)dt, \tag{10.6}$$

where $|| \cdot ||$ denotes the 2-norm of a vector, $T > 0$ denotes a bounded final time, and J is a strictly increasing function of its argument (e.g., an energy consumption function $J(||u(t)||) = ||u(t)||^2$). We consider the following additional requirements:

Trajectory tracking: We require the vehicle to stay as close as possible to a desired *reference trajectory* X_r (e.g., the middle of its current lane).

State constraints: We impose a set of constraints (componentwise) on the state of system (2.11) in the following form:

$$x_{min} \leq x(t) \leq x_{max}, \forall t \in [0, T], \tag{10.7}$$

where $x_{max} := (x_{max,1}, x_{max,2}, \ldots, x_{max,n}) \in \mathbb{R}^n$ and $x_{min} := (x_{min,1}, x_{min,2}, \ldots, x_{min,n}) \in \mathbb{R}^n$ denote the maximum and minimum state vectors, respectively. Examples of such constraints for a vehicle include maximum acceleration, maximum braking, and maximum steering rate.

Priority structure: We require the system trajectory X of (2.11) starting at $x(0) = x_0$ to satisfy a priority structure $\langle R, \sim_p, \leq_p \rangle$, i.e.,

$$X \models \langle R, \sim_p, \leq_p \rangle, \tag{10.8}$$

where $\sim_p$ is an equivalence relation over a finite set of rules R and $\leq_p$ is a total order over the equivalence classes. Our priority structure, called **Total ORder over eQuivalence classes (TORQ)**, is related to the rulebooks from [13]. However, rather than allowing for a partial order over the set of rules R, we require that any two rules are either comparable or equivalent. Informally, (10.8) means that X is the "best" trajectory that (2.11) can produce, considering the violation metrics of the rules in R and the priorities captured by $\sim_p$ and $\leq_p$. A formal definition for a priority structure and its satisfaction will be given in the sequel.

Control bounds: We impose control bounds as given in (2.19). Examples include jerk and steering acceleration.

Formally, we can define the optimal control problem with rules (**OCP-R**) as follows:

OCP-R: Find a control policy for system (2.11) such that the objective function in (10.6) is minimized, and the trajectory tracking, state constraints (10.7), the TORQ priority structure $\langle R, \sim_p, \leq_p \rangle$, and control bounds (2.19) are satisfied by the generated trajectory given $\boldsymbol{x}(0)$. $\qquad\qquad\qquad\square$

Our approach to Problem **OCP-R** can be summarized as follows. We use CLFs for tracking the reference trajectory $\mathcal{X}_r$ and HOCBFs to implement the state constraints (10.7). For each rule in R, we define violation metrics. We show that satisfaction of the rules can be expressed as forward invariance for sets described by differentiable functions, and enforce them using HOCBFs. The control bounds (2.19) are considered as constraints. We provide an iterative solution to Problem **OCP-R**, where each iteration involves solving a sequence of QPs. In the first iteration, all the rules from R are considered. If the corresponding QPs are feasible, then an optimal control is found. Otherwise, we iteratively relax the satisfaction of rules from subsets of R based on their priorities, and minimize the corresponding relaxations by including them in the cost function.

10.2.2 Rules and Priority Structures

In this section, we extend the rulebooks from [13] by formalizing the rules and defining violation metrics. We introduce the TORQ priority structure from the last section, in which all rules are comparable, and it is particularly suited for the hierarchical optimal control framework shown in the sequel.

Rules. In the definition below, an *instance* $i \in S_p$ is a traffic participant or artifact that is involved in a rule, where S_p is the set of all instances involved in the rule. For example, in a rule to maintain clearance from pedestrians, a pedestrian is an instance, and there can be many instances encountered by ego in a given scenario. Instances can also be traffic artifacts like the road boundary (of which there is only one), lane boundaries, or stop lines.

> **Definition 10.1** (*Rule*) A rule is composed of a statement and three violation metrics. A statement is a formula that is required to be satisfied for all times. A formula is inductively defined as:
>
> $$\varphi := \mu \,|\, \neg\varphi \,|\, \varphi_1 \wedge \varphi_2, \qquad\qquad (10.9)$$
>
> where $\varphi, \varphi_1, \varphi_2$ are formulas, $\mu := (h(\boldsymbol{x}) \geq 0)$ is a predicate on the state vector $\boldsymbol{x}$ of system (2.11) with $h : \mathbb{R}^n \to \mathbb{R}$. $\wedge, \neg$ are Boolean operators for conjunction and negation, respectively. The three violation metrics for a rule r are defined as:

1. Instantaneous violation metric $\varrho_{r,i}(\boldsymbol{x}(t)) \in [0, 1]$,
2. Instance violation metric $\rho_{r,i}(\mathcal{X}) \in [0, 1]$, and
3. Total violation metric $P_r(\mathcal{X}) \in [0, 1]$,

where i is an instance, $\boldsymbol{x}(t)$ is a trajectory at time t and $\mathcal{X}$ is a whole trajectory of ego. The instantaneous violation metric $\varrho_{r,i}(\boldsymbol{x}(t))$ quantifies violation by a trajectory at a specific time t with respect to a given instance i. The instance violation metric $\rho_{r,i}(\mathcal{X})$ captures a violation with respect to a given instance i over the whole time of a trajectory; it is obtained by aggregating $\varrho_{r,i}(\boldsymbol{x}(t))$ over the entire time of a trajectory $\mathcal{X}$. The total violation metric P_r is the aggregation of the instance violation metric $\rho_{r,i}(\mathcal{X})$ over all instances $i \in S_p$.

The aggregations in the above definition can be implemented through a selection of a maximum or a minimum, integration over time, summation over instances, or by using general L_p norms. A zero value for a violation score shows satisfaction of the rule. A strictly positive value denotes violation—the larger the score, the more the ego AV violates the rule. In what follows, for simplicity, we use ϱ_r and ρ_r instead of $\varrho_{r,i}$ and $\rho_{r,i}$ if there is only one instance. Examples of rules (statements and violations metrics and scores) are given in the simulation examples in Sect. 10.2.4.

We divide the set of rules into two categories: (*i*) *clearance rules*—safety relevant rules enforcing the requirement that the ego AV maintains a minimal distance to other traffic participants and to the side of the road or lane, and (*ii*) *non-clearance rules*—rules that are not contained in the first category, such as speed limits. In the following, we provide a general methodology to express clearance rules as inequalities involving differentiable functions, which will allow us to enforce their satisfaction using HOCBFs.

Remark 10.1 The violation metrics from Definition 10.1 are inspired from Signal Temporal Logic (STL) robustness [18, 37, 39], which quantifies how a signal (trajectory) satisfies a temporal logic formula. Here, we focus on rules that we aim to satisfy at all times. Therefore, the rules in (10.9) can be seen as (particular) STL formulas, which all start with an "always" temporal operator (omitted here).

Priority Structure. The pre-order rulebook from [13] defines a "base" pre-order that captures relative priorities of some (comparable) rules, which are often similar in different states and countries. A pre-order rulebook can be made more precise for a specific legislation by adding rules and/or priority relations through priority refinement, rule aggregation and augmentation. This can be done through empirical studies or learning from local data to construct a total order rulebook. To order trajectories, authors of [13] enumerated all the total orders compatible with a given pre-order. Motivated by the hierarchical optimal control framework, we require that any two rules are in a relationship, in the sense that they are either equivalent or comparable with respect to their priorities.

> **Definition 10.2** (*TORQ Priority Structure*) A Total ORder over eQuivalence classes (TORQ) priority structure is a tuple $\langle R, \sim_p, \leq_p \rangle$, where R is a finite set of rules, $\sim_p$ is an equivalence relation over R, and $\leq_p$ is a total order over the set of equivalence classes determined by $\sim_p$.

Equivalent rules (i.e., rules in the same class) have the same priority. Given two equivalence classes O_1 and O_2 with $O_1 \leq_p O_2$, every rule $r_1 \in O_1$ has lower priority than every rule $r_2 \in O_2$. Since $\leq_p$ is a total order, any two rules $r_1, r_2 \in R$ are comparable, in the sense that exactly one of the following three statements is true: (i) r_1 and r_2 have the same priority, (ii) r_1 has higher priority than r_2, and (iii) r_2 has higher priority than r_1.

Given a TORQ $\langle R, \sim_p, \leq_p \rangle$, we can assign numerical (integer) priorities to the rules. We assign priority 1 to the equivalence class with the lowest priority, priority 2 to the next one and so on. The rules inside an equivalence class inherit the priority from their equivalence class. Given a priority structure $\langle R, \sim_p, \leq_p \rangle$ and violation scores for the rules in R, we can compare trajectories:

> **Definition 10.3** (*Trajectory Comparison*) A trajectory X_1 is said to be **better** (less violating) than another trajectory X_2 if the highest priority rule(s) violated by X_1 has a lower priority than the highest priority rule(s) violated by X_2. If both trajectories violate an equivalent highest priority rule(s), then the one with the smaller (maximum) total violation score is better. In this case, if the trajectories have equal violation scores, then they are equivalent.

It is easy to see that, by following Definition 10.3, given two trajectories, one can be better than the other, or they can be equivalent (i.e., two trajectories cannot be incomparable).

Example 10.1 Consider the driving scenario from Fig. 10.3a and TORQ $\langle R, \sim_p, \leq_p \rangle$ in Fig. 10.3b, where $R = \{r_1, r_2, r_3, r_4\}$, and r_1: "No collision," r_2: "Lane keeping," r_3: "Speed limit" and r_4: "Comfort". There are three equivalence classes given by $O_1 = \{r_4\}$, $O_2 = \{r_2, r_3\}$ and $O_3 = \{r_1\}$. Rule r_4 has priority 1, r_2 and r_3 have priority 2, and r_1 has priority 3. Assume the instance (same as total, as there is only one instance for each rule) violation scores of rules $r_i, i = 1, 2, 3, 4$ by trajectories a, b, c are given by $\rho_i = (\rho_i(a), \rho_i(b), \rho_i(c))$ as shown in Fig. 10.3b. Based on Definition 10.3, trajectory b is better (less violating) than trajectory a since the highest priority rule violated by b (r_2) has a lower priority than the highest priority rule violated by a (r_1). The same argument holds for trajectories a and c, i.e., c is better than a. The highest priority rules violated by trajectories b and c have the same

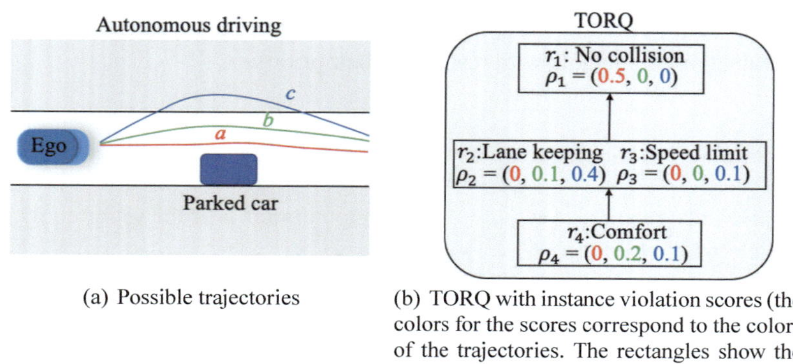

Autonomous driving

(a) Possible trajectories

(b) TORQ with instance violation scores (the colors for the scores correspond to the colors of the trajectories. The rectangles show the equivalence classes.

Fig. 10.3 An autonomous driving scenario with three possible trajectories, 4 rules, and 3 equivalence classes

priorities. Since the maximum violation score of the highest priority rules violated by b is smaller than that for c, i.e., $\max(\rho_2(b), \rho_3(b)) = 0.1$, $\max(\rho_2(c), \rho_3(c)) = 0.4$, trajectory b is better than c. ∎

> **Definition 10.4** (*TORQ satisfaction*) A trajectory X of system (2.11) starting at $x(0)$ satisfies a TORQ $\langle R, \sim_p, \leq_p \rangle$ (i.e., $X \models \langle R, \sim_p, \leq_p \rangle$), if there are no better trajectories of (2.11) starting at $x(0)$.

Definition 10.4 is central to our solution of Problem **OCP-R**, which is based on an iterative relaxation of the rules according to their satisfaction of the TORQ.

10.2.3 Rule-Based Optimal Control

In this section, we present the detailed steps of our approach to solve Problem **OCP-R**.

10.2.3.1 Trajectory Tracking

The system dynamics in (2.11) are viewed as "traditional" vehicle dynamics with respect to an inertial reference frame [4] or as dynamics defined along a given reference trajectory [53] (see (10.21)). The problem setting considered here falls in the second category (the middle of ego's current lane is the default reference trajectory). We use the model from [53], in which part of the state of (2.11) captures the tracking errors with respect to the reference trajectory. The tracking problem then becomes one of stabilizing the error states to 0. Suppose the error state vector is $y \in R^{n_0}$, $n_0 \leq n$ (the components in y are a subset of the components in x). We define a CLF $V(x) = ||y||^2$. Any control u that satisfies the relaxed CLF constraint [4] given by

$$L_f V(x) + L_g V(x)u + \epsilon V(x) \leq \delta_e, \tag{10.10}$$

exponentially stabilizes the error states to 0 if $\delta_e(t) = 0$, $\forall t \in [0, T]$, where $\delta_e > 0$ is a relaxation variable that compromises between stabilization and feasibility. Note that the CLF constraint (10.10) only works for $V(x)$ with relative degree 1. If the relative degree is larger than 1, we can use input-to-state linearization and state feedback control [26] to reduce the relative degree to one [77].

10.2.3.2 Clearance and Optimal Disk Coverage

Satisfaction of a priority structure can be enforced by formulating real-time constraints on the ego AV state $x(t)$ that appear in the violation metrics. Satisfaction of the non-clearance rules can be easily implemented using HOCBFs. For clearance rules, we define a notion of clearance region around ego and around the traffic participants in S_p that are involved in the rule (e.g., pedestrians and other vehicles).

The clearance region for the ego AV is defined as a rectangle with tunable speed-dependent lengths (i.e., we may choose to have a larger clearance from pedestrians when the ego AV is driving at higher speeds) and defined based on the ego AV's footprint and functions $h_f(x)$, $h_b(x)$, $h_l(x)$, $h_r(x)$ that determine the front, back, left, and right clearances as illustrated in Fig. 10.4, where $h_f, h_b, h_l, h_r : \mathbb{R}^n \to \mathbb{R}_{\geq 0}$. The clearance regions for participants (instances) are defined such that they comply with their geometry and cover their footprints, e.g., (fixed-length) rectangles for other vehicles and (fixed-radius) disks for pedestrians, as shown in Fig. 10.4.

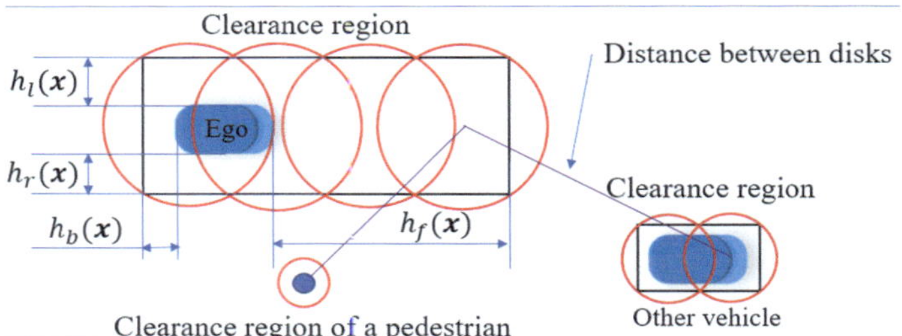

Fig. 10.4 The clearance regions and their coverage with disks: the clearance region and the disks are speed-dependent for the ego AV and fixed for the other vehicle and the pedestrian. We consider the distances between all the possible pairs of disks from the ego AV and other traffic participants (e.g., pedestrians, parked vehicles). There are 12 distance pairs in total, and we only show two of them w.r.t. the pedestrian and another vehicle, respectively

To satisfy a clearance rule involving traffic participants, we need to avoid any overlaps between the clearance regions of the ego AV and its traffic participants. We define a function $d_{min}(\boldsymbol{x}, \boldsymbol{x}_i) : \mathbb{R}^{n+n_i} \to \mathbb{R}$ to determine the signed distance between the clearance regions of the ego AV and participant $i \in S_p$ ($\boldsymbol{x}_i \in \mathbb{R}^{n_i}$ denotes the state of participant i), which is negative if the clearance regions overlap. Therefore, satisfaction of a clearance rule can be imposed by having a constraint on $d_{min}(\boldsymbol{x}, \boldsymbol{x}_i)$ be non-negative. For the clearance rules "stay in lane" and "stay in drivable area," we require that the ego AV clearance region be within the lane and the drivable area, respectively.

However, finding $d_{min}(\boldsymbol{x}, \boldsymbol{x}_i)$ can be computationally expensive. For example, the distance between two rectangles could be from corner to corner, corner to edge, or edge to edge. Since each rectangle has four corners and four edges, there are 64 possible cases. More importantly, this computation leads to a non-smooth $d_{min}(\boldsymbol{x}, \boldsymbol{x}_i)$ function, which cannot be used to enforce clearance using a CBF approach. To address these issues, we propose an optimal coverage of the rectangles with disks, which allows to map the satisfaction of the clearance rules to a set of smooth HOCBF constraints (i.e., there will be one constraint for each pair of centers of disks pertaining to different traffic participants).

We use $l > 0$ and $w > 0$ to denote the length and width of the ego AV's footprint, respectively. Assume we use $z \in \mathbb{N}$ disks with centers located on the center line of the clearance region to cover it (see Fig. 10.5). Since all the disks have the same radius, the minimum radius to fully cover the ego AV's clearance region, denoted by $\mathfrak{r} > 0$, is given by:

$$\mathfrak{r} = \sqrt{\left(\frac{w + h_l(\boldsymbol{x}) + h_r(\boldsymbol{x})}{2}\right)^2 + \left(\frac{l + h_f(\boldsymbol{x}) + h_b(\boldsymbol{x})}{2z}\right)^2}. \tag{10.11}$$

The minimum radius $\mathfrak{r}_i$ of the rectangular clearance region for a traffic participant $i \in S_p$ with disks number z_i is defined in a similar way using the length and width of its footprint and setting $h_l, h_r, h_b, h_f = 0$.

Assume the center of the disk $j \in \{1, \ldots, z\}$ for the ego AV and the center of the disk $k \in \{1, \ldots, z_i\}$ for the instance $i \in S_p$ are given by $(x_{e,j}, y_{e,j}) \in \mathbb{R}^2$ and $(x_{i,k}, y_{i,k}) \in \mathbb{R}^2$, respectively (see Fig. 10.5). To avoid any overlap between the corresponding disks of ego and the instance $i \in S_p$, we impose the following constraints:

Fig. 10.5 The optimal disk coverage of a clearance region

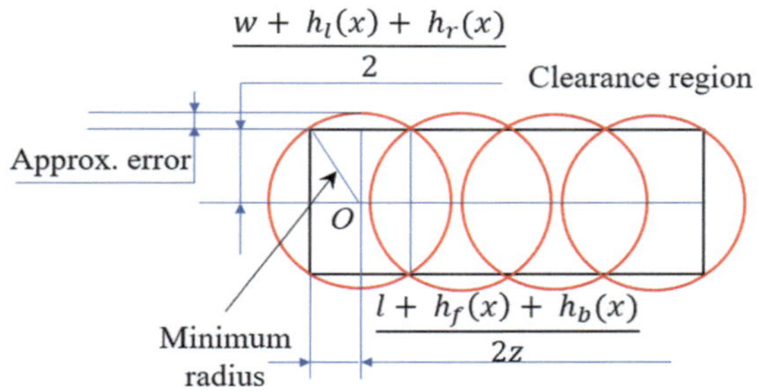

$$\sqrt{(x_{e,j} - x_{i,k})^2 + (y_{e,j} - y_{i,k})^2} \geq \mathfrak{r} + \mathfrak{r}_i,$$

$$\forall j \in \{1, \ldots, z\}, \forall k \in \{1, \ldots, z_i\}. \tag{10.12}$$

Since disks fully cover the clearance regions, enforcing (10.12) also guarantees that $d_{min}(\boldsymbol{x}, \boldsymbol{x}_i) \geq 0$. For the clearance rules "stay in lane" and "stay in drivable area," we can get similar constraints as (10.12) to make the disks that cover the ego AV's clearance region stay within them (e.g., we can consider $h_l, h_r, h_b, h_f = 0$ and formulate (10.12) such that the distance between the ego AV disk centers and the line in the middle of its current lane be less than $\frac{w_l}{2} - \mathfrak{r}$, where $w_l > 0$ denotes the lane width). Thus, we can formulate satisfaction of all the clearance rules as continuously differentiable constraints (10.12), and implement them using HOCBFs.

To efficiently formulate the proposed optimal disk coverage approach, we need to find the minimum number of the disks that fully cover the clearance regions as it determines the number of constraints in (10.12). Moreover, we need to minimize the lateral approximation error since large errors imply overly conservative constraint (see Fig. 10.5). This can be formally defined as an optimization problem, and solved offline to determine the numbers and radii of the disks in (10.12). In what follows, we present this procedure.

To construct disks to fully cover the clearance regions, we need to find their number and radius. From Fig. 10.5, the lateral approximation error $\sigma > 0$ is given by:

$$\sigma = \mathfrak{r} - \frac{w + h_l(\boldsymbol{x}) + h_r(\boldsymbol{x})}{2}. \tag{10.13}$$

Since σ for the ego AV depends on its state $\boldsymbol{x}$ (speed-dependent), we consider the accumulated lateral approximation error for all possible $\boldsymbol{x} \in X$. This allows us to determine z and $\mathfrak{r}$ such that the disks fully cover the ego AV clearance region for all possible speeds in $\boldsymbol{x}$. Let $\bar{h}_i = \sup_{\boldsymbol{x} \in X} h_i(\boldsymbol{x}), \underline{h}_i = \inf_{\boldsymbol{x} \in X} h_i(\boldsymbol{x}), i \in \{f, b, l, r\}$. We can formally formulate the construction of the approximation disks as an optimization problem:

$$\min_z z + \beta \int_{\underline{h}_f}^{\bar{h}_f} \int_{\underline{h}_b}^{\bar{h}_b} \int_{\underline{h}_l}^{\bar{h}_l} \int_{\underline{h}_r}^{\bar{h}_r} \sigma \, dh_f(\boldsymbol{x}) dh_b(\boldsymbol{x}) dh_l(\boldsymbol{x}) dh_r(\boldsymbol{x}) \tag{10.14}$$

subject to

$$z \in \mathbb{N}, \tag{10.15}$$

where $\beta \geq 0$ is a trade-off between minimizing the number of the disks (so as to minimize the number of constraints considered with CBFs) and the coverage approximation error. The above optimization problem is solved offline. A similar optimization problem is formulated for the construction of disks for instances in S_p (we remove the integrals due to speed independence). Note that for the driving scenarios studied in this chapter, we omit the longitudinal approximation errors in the front and back. The lateral approximation errors are considered in the disk formulation since they induce conservativeness in the lateral maneuvers of the ego AV required for surpassing other instances (such as parked car, pedestrians, etc.); see Sect. 10.2.4.

Let $(x_e, y_e) \in \mathbb{R}^2$ be the center of the ego AV and $(x_i, y_i) \in \mathbb{R}^2$ be the center of instance $i \in S_p$. The center of disk j for the ego AV $(x_{e,j}, y_{e,j})$, $j \in \{1, \ldots, z\}$ is determined by:

$$
\begin{aligned}
x_{e,j} &= x_e + \cos\theta_e \left(-\frac{l}{2} - h_b(\mathbf{x}) + \frac{l + h_f(\mathbf{x}) + h_b(\mathbf{x})}{2z}(2j - 1) \right) \\
y_{e,j} &= y_e + \sin\theta_e \left(-\frac{l}{2} - h_b(\mathbf{x}) + \frac{l + h_f(\mathbf{x}) + h_b(\mathbf{x})}{2z}(2j - 1) \right)
\end{aligned}
\tag{10.16}
$$

where $j \in \{1, \ldots, z\}$ and $\theta_e \in \mathbb{R}$ denotes the heading angle of the ego AV. The center of disk k for instance $i \in S_p$ denoted by $(x_{i,k}, y_{i,k})$, $k \in \{1, \ldots, z_i\}$, can be defined similarly.

Theorem 10.1 *If the clearance regions of the ego AV and instance $i \in S_p$ are covered by the disks constructed by solving (10.14), then the clearance regions of ego and instance i do not overlap if (10.12) is satisfied.*

Proof Let z and z_i be the disks with minimum radius $\mathfrak{r}$ and $\mathfrak{r}_i$ from (10.11) associated with the clearance regions of the ego AV and instance $i \in S_p$, respectively. The constraints in (10.12) guarantee that there is no overlap of the disks between vehicle $i \in S_p$ and instance $j \in S_p$. Since the clearance regions are fully covered by these disks, we conclude that the clearance regions do not overlap. $\qquad\square$

10.2.3.3 Optimal Control

We now present our complete framework to solve problem **OCP-R**. We propose a recursive algorithm to iteratively relax the satisfaction of the rules in the priority structure $\langle R, \sim_p, \leq_p \rangle$ (if needed) based on the total order over the equivalence classes.

Let R_O be the set of equivalence classes in $\langle R, \sim_p, \leq_p \rangle$, and N_O be the cardinality of R_O. We construct the power set of equivalence classes denoted by $S = 2^{R_O}$, and incrementally (from low to high priority) sort the sets in S based on the highest priority of the equivalence classes in each set according to the total order and denote the sorted set by $S_{sorted} = \{S_1, S_2, \ldots, S_{2^{N_O}}\}$, where $S_1 = \{\emptyset\}$. We use this sorted set in our optimal control formulation to obtain satisfaction of the higher priority classes, even at the cost of relaxing satisfaction of the lower priority classes. Therefore, from Definition 10.4, the solution of the optimal control will satisfy the priority structure.

Example 10.2 (*Example* 10.1 *revisited*). We define $R_O = \{O_1, O_2, O_3\}$. Based on the given total order $O_1 \leq_p O_2 \leq_p O_3$, we can write the sorted power set as $S_{sorted} = \{\{\emptyset\}, \{O_1\}, \{O_2\}, \{O_1, O_2\}, \{O_3\}, \{O_1, O_3\}, \{O_2, O_3\}, \{O_1, O_2, O_3\}\}$. ∎

In order to find a trajectory that satisfies a given TORQ, we first assume that all the rules are satisfied. Starting from $S_1 = \{\emptyset\}$ in the sorted set S_{sorted}, we solve Problem **OCP-R** given that no rules are relaxed, i.e., all the rules must be satisfied. If the problem is infeasible, we move to the next set $S_2 \in S_{sorted}$, and relax all the rules of all the equivalence classes in S_2 while enforcing satisfaction of all the other rules in the equivalence class set denoted by $R_O \setminus S_2$. This procedure is done recursively until we find a feasible solution of Problem **OCP-R**. Formally, at $k = 1, 2 \ldots, 2^{N_O}$ for $S_k \in S_{sorted}$, we relax all the rules $i \in O$ for all the equivalence classes $O \in S_k$ and reformulate Problem **OCP-R** as the following optimal control problem:

$$\min_{u, \delta_e, \delta_i, i \in O, O \in S_k} \int_0^T \left[J(\|u\|) + p_e \delta_e^2 + \sum_{i \in O, O \in S_k} p_i \delta_i^2 \right] dt \tag{10.17}$$

subject to: dynamics (2.11), control bounds (2.19), CLF constraint (10.10), and

$$L_f^{m_j} b_j(x) + L_g L_f^{m_j - 1} b_j(x) u + O(b_j(x))$$
$$+ \alpha_{m_j}(\psi_{m_j - 1}(x)) \geq 0, \ \forall j \in O, \forall O \in R_O \setminus S_k, \tag{10.18}$$

$$L_f^{m_i} b_i(x) + L_g L_f^{m_i - 1} b_i(x) u + O(b_i(x))$$
$$+ \alpha_{m_i}(\psi_{m_i - 1}(x)) \geq \delta_i, \ \forall i \in O, \forall O \in S_k, \tag{10.19}$$

$$L_f^{m_l} b_l(x) + L_g L_f^{m_l - 1} b_{lim,l}(x) u + O(b_{lim,l}(x))$$
$$+ \alpha_{m_l}(\psi_{m_l - 1}(x)) \geq 0, \ \forall l \in \{1, \ldots, 2n\}, \tag{10.20}$$

where $p_e > 0$ and $p_i > 0, i \in O, O \in S_k$ assign the trade-off between the CLF relaxation δ_e (used for trajectory tracking) and the HOCBF relaxations δ_i. m_i, m_j, m_l denotes the relative degree of $b_i(x), b_j(x), b_{lim,l}(x)$, respectively. The functions $b_i(x)$ and $b_j(x)$ are HOCBFs for the rules in $\langle R, \sim_p, \leq_p \rangle$, and are implemented directly from the rule statement for non-clearance rules or by using the optimal disk coverage framework for clearance rules. At relaxation step k, HOCBFs corresponding to the rules in $O, \forall O \in S_k$ are relaxed by adding $p_i > 0, i \in O, O \in S_k$ in (10.19), while for other rules in R in (10.18) and the state constraints (10.20), regular HOCBFs are used. We assign $p_i, i \in O, O \in S_k$ according to their relative priorities, i.e., we choose a larger p_i for the rule i that belongs to a higher priority class. The functions $b_{lim,l}(x), l \in \{1, \ldots, 2n\}$ are HOCBFs for the state limitations (10.7). The functions $\psi_{m_i}(x), \psi_{m_j}(x), \psi_{m_l}(x)$ are defined as in (3.3). $\alpha_{m_i}, \alpha_{m_j}, \alpha_{m_l}$ can be penalized to improve the feasibility of the problem above [71, 74].

If the above optimization problem is feasible for all $t \in [0, T]$, we can specifically determine which rules (within an equivalence class) are relaxed based on the values of

δ_i, $i \in O$, $O \in S_k$ in the optimal solution (i.e., if $\delta_i(t) = 0$, $\forall t \in \{0, T\}$, then rule i does not need to be relaxed). This procedure is summarized in Algorithm 5.

Remark 10.2 (*Complexity*) The optimization problem (10.17) is solved using the usual sequence of QPs introduced in Chap. 2. The complexity of each QP is $O(y^3)$, where $y \in \mathbb{N}$ is the dimension of decision variables. It usually takes less than $0.01s$ to solve each QP in MATLAB. The total time for each iteration $k \in \{1, \ldots, 2^{N_O}\}$ depends on the final time T and the length of the reference trajectory X_r. The computation time can be further improved by running the code in parallel over multiple processors.

10.2.3.4 Pass/Fail Evaluation

As an extension to Problem **OCP-R**, we formulate and solve a pass/fail (P/F) procedure, in which we are given a vehicle trajectory, and the goal is to accept (pass, P) or reject (fail, F) it based on the satisfaction of the rules. Specifically, given a candidate trajectory X_c of system (2.11), and given a TORQ $\langle R, \sim_p, \leq_p \rangle$, we pass (P) X_c if we cannot find a better trajectory according to Definition 10.3. Otherwise, we fail (F) X_c. We proceed as follows: We find the total violation scores of the rules in $\langle R, \sim_p, \leq_p \rangle$ for the candidate trajectory X_c. If no rules in R are violated, then we pass the candidate trajectory. Otherwise, we investigate the existence of a better (less violating) trajectory. We take the middle of ego's current lane as the reference trajectory X_r and re-formulate the optimal control problem in (10.17) to recursively relax rules such that if the optimization is feasible, the generated trajectory is better than the candidate trajectory X_c. Specifically, assume that the highest priority rule(s) that the candidate trajectory X_c violates belongs to O_H, $H \in \mathbb{N}$. Let $R_H \subseteq R_O$ denote the set of equivalence classes with priorities not larger than H, and $N_H \in \mathbb{N}$ denote the cardinality of R_H. We construct a power set $S_H = 2^{R_H}$, and then apply Algorithm 5, in which we replace R_O by R_H.

Remark 10.3 The procedure described above would fail a candidate trajectory X_c even if only a slightly better alternate trajectory (i.e., violating rules of the same highest priority but with slightly smaller violation scores) can be found by solving the optimal control problem. In practice, this might lead to an undesirably high failure rate. One way to deal with this is to allow for more classification categories, e.g., "Provisional Pass" (PP), which can then trigger further investigation of X_c.

Example 10.3 (*Example* 10.1 *revisited*) Returning to the example in Fig. 10.3, assume trajectory b is a candidate trajectory which violates rules r_2, r_4, thus, the highest priority rule that is violated by trajectory b belongs to O_2. We construct $R_H = \{O_1, O_2\}$. The power set $S_H = 2^{R_H}$ is then defined as $S_H = \{\{\emptyset\}, \{O_1\}, \{O_2\}, \{O_1, O_2\}\}$, and is sorted based on the total order as $S_{H_{sorted}} = \{\{\emptyset\}, \{O_1\}, \{O_2\}, \{O_1, O_2\}\}$.   ∎

Algorithm 5: Recursive relaxation algorithm for finding optimal trajectory

---

Input: System (2.11) with $x(0)$, cost function (10.6), control bound (2.19), state constraint
(10.7), TORQ $\langle R, \sim_p, \leq_p \rangle$, reference trajectory X_r

Output: Optimal ego trajectory and set of relaxed rules

1. Construct the power set of equivalence classes $S = 2^{R_O}$;

2. Sort the sets in S based on the highest priority of the equivalence classes in each set
according to the total order and get $S_{sorted} = \{S_1, S_2, \ldots, S_{2^{N_O}}\}$;

3. $k = 0$;

while $k + + \leq 2^{N_O}$ **do**

    Solve (10.17) s.t. (2.11), (2.19), (10.10), (10.19), (10.18) and (10.20);

    if *the above problem is feasible for all* $t \in [0, T]$ **then**

        Generate the optimal trajectory X^* from (2.11);

        Construct relaxed set $R_{relax} = \{i : i \in O, O \in S_k\}$;

        if $\delta_i(t) = 0, \forall t \in [0, T]$ **then**

            Remove i from R_{relax};

        end

        break;

    end

end

4. Return X^* and R_{relax};

10.2.4 Simulation Examples

In this section, we apply the methodology we have developed to specific vehicle dynamics
and various driving scenarios. Ego AV dynamics (2.11) are defined with respect to a reference
trajectory [53], which measures the along-trajectory distance $s \in \mathbb{R}$ and the lateral distance
$d \in \mathbb{R}$ of the vehicle Center of Gravity (CoG) with respect to the closest point on the reference
trajectory as follows:

$$
\underbrace{\begin{bmatrix} \dot{s} \\ \dot{d} \\ \dot{\mu} \\ \dot{v} \\ \dot{a} \\ \dot{\delta} \\ \dot{\omega} \end{bmatrix}}_{\dot{x}} = \underbrace{\begin{bmatrix} \frac{v\cos(\mu+\beta)}{1-d\kappa} \\ v\sin(\mu+\beta) \\ \frac{v}{l_r}\sin\beta - \kappa\frac{v\cos(\mu+\beta)}{1-d\kappa} \\ a \\ 0 \\ \omega \\ 0 \end{bmatrix}}_{f(x)} + \underbrace{\begin{bmatrix} 0 & 0 \\ 0 & 0 \\ 0 & 0 \\ 0 & 0 \\ 1 & 0 \\ 0 & 0 \\ 0 & 1 \end{bmatrix}}_{g(x)} \underbrace{\begin{bmatrix} u_{jerk} \\ u_{steer} \end{bmatrix}}_{u}, \tag{10.21}
$$

where μ is the vehicle local heading error determined by the difference of the global vehicle
heading $\theta \in \mathbb{R}$ in (10.16) and the tangent angle $\phi \in \mathbb{R}$ of the closest point on the reference
trajectory (i.e., $\theta = \phi + \mu$); v, a denote the vehicle linear speed and acceleration; δ, ω
denote the steering angle and steering rate, respectively; κ is the curvature of the reference

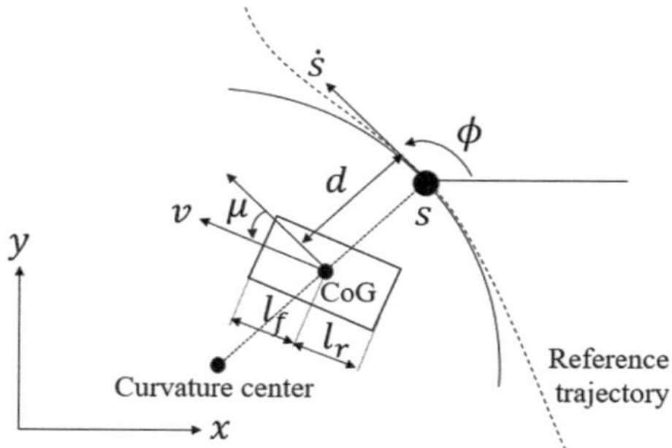

Fig. 10.6 Coordinates of ego AV with respect to a reference trajectory

trajectory at the closest point; l_r is the length of the vehicle from the tail to the CoG; and u_{jerk}, u_{steer} denote the two control inputs for jerk and steering acceleration as shown in Fig. 10.6. $\beta = \arctan\left(\frac{l_r}{l_r+l_f}\tan\delta\right)$ where l_f is the length of the vehicle from the head to the CoG.

We consider the cost function in (10.17) as:

$$\min_{u_{jerk}(t), u_{steer}(t)} \int_0^T \left[u_{jerk}^2(t) + u_{steer}^2(t)\right] dt. \tag{10.22}$$

The reference trajectory $\mathcal{X}_r$ is the middle of ego's current lane, and is assumed to be given as an ordered sequence of points $p_1, p_2, \ldots, p_{N_r}$, where $p_i \in \mathbb{R}^2, i = 1, \ldots, N_r$ (N_r denotes the number of points). We can find the reference point $p_{i(t)}, i : [0, T] \to \{1, \ldots, N_r\}$ at time t as follows:

$$i(t) = \begin{cases} i(t) + 1 & ||p(t) - p_{i(t)}|| \le \gamma, \\ j & \exists j \in \{1, 2, \ldots, N_r\} : ||p(t) - p_{i(t)}|| \ge ||p(t) - p_j||, \end{cases} \tag{10.23}$$

where $p(t) \in \mathbb{R}^2$ denotes ego's location. $\gamma > 0$, and $i(0) = k$ for a $k \in \{1, 2, \ldots, N_r\}$ is chosen such that $||p(0) - p_j|| \ge ||p(0) - p_k||, \forall j \in \{1, 2, N_r\}$. Once we get $p_{i(t)}$, we can update the progress s, the error states d, μ and the curvature κ in (10.21). The trajectory tracking in this case is to stabilize the error states d, μ ($y = (d, \mu)$ in (10.10)) to 0. We also wish the ego AV to achieve a desired speed $v_d > 0$ (otherwise, it may stop in curved lanes). We achieve this by re-defining the CLF $V(x)$ in (10.10) as $V(x) = ||y||^2 + c_0(v - v_d)^2, c_0 > 0$. As the relative degree of $V(x)$ w.r.t. (10.21) is larger than 1, we use input-to-state linearization and state feedback control [26] to reduce the relative degree to one [77]. For example, for the desired speed part in the CLF $V(x)$ ((10.21) is in linear form from v to u_{jerk}, so we do not need to do linearization), we can find a desired state feedback acceleration $\hat{a} = -k_1(v - v_d), k_1 > 0$. Then we can define a new CLF in the form $V(x) = ||y||^2 + c_0(a - \hat{a})^2 = ||y||^2 + c_0(a + k_1(v - v_d))^2$ whose relative degree is just one w.r.t.

u_{jerk} in (10.21). We proceed similarly for driving d, μ to 0 in the CLF $V(x)$ as the relative degrees of d, μ are also larger than one.

The control bounds (2.19) and state constraints (10.7) are given by:

$$
\begin{aligned}
\text{speed constraint:} \quad & v_{\min} \leq v(t) \leq v_{\max}, \\
\text{acceleration constraint:} \quad & a_{\min} \leq a(t) \leq a_{\max}, \\
\text{jerk control constraint:} \quad & u_{j,\min} \leq u_{jerk}(t) \leq u_{j,\max}, \\
\text{steering angle constraint:} \quad & \delta_{\min} \leq \delta(t) \leq \delta_{\max}, \\
\text{steering rate constraint:} \quad & \omega_{\min} \leq \omega(t) \leq \omega_{\max}, \\
\text{steering control constraint:} \quad & u_{s,\min} \leq u_{steer}(t) \leq u_{s,\max},
\end{aligned}
\tag{10.24}
$$

We use a set of 8 rules, which we describe below. According to Definition 10.1, each rule statement should be satisfied at all times.

r_1 : Maintain clearance with pedestrians

Statement: $d_{min,fp}(x, x_i) \geq d_1 + v(t)\eta_1, \forall i \in S_{ped}$

$$
\varrho_{r,i}(x(t)) = \max\left(0, \frac{d_1 + v(t)\eta_1 - d_{min,fp}(x, x_i)}{d_1 + v_{max}\eta_1}\right)^2,
\tag{10.25}
$$

$$
\rho_{r,i}(X) = \max_{t \in [0,T]} \varrho_{r,i}(x(t)), \quad P_r = \sqrt{\frac{1}{n_{ped}} \sum_{i \in S_{ped}} \rho_{r,i}}.
$$

where $d_{min,fp} : \mathbb{R}^{n+n_i} \to \mathbb{R}$ denotes the distance between the footprints of the ego AV and pedestrian i, and $d_{min,fp}(\cdot, \cdot) < 0$ denotes the footprint overlap. The clearance threshold is given based on a fixed distance $d_1 \geq 0$ and increases linearly by $\eta_1 > 0$ based on ego AV speed $v(t) \geq 0$ (d_1 and η_1 are determined empirically). S_{ped} denotes the index set of all pedestrians, and $x_i \in \mathbb{R}^{n_i}$ denotes the state of pedestrian i. v_{max} is the maximum feasible speed of the vehicle and is used to define the normalization term in $\varrho_{r,i}$, which assigns a violation score (based on a L-2 norm) if formula is violated by $x(t)$. $\rho_{r,i}$ defines the instance violation score as the most violating instant over X. P_r aggregates the instance violations over all units (pedestrians), where $n_{ped} \in \mathbb{N}$ denotes the number of pedestrians.

r_2 : Stay in the drivable area

Statement: $d_{left}(x(t)) + d_{right}(x(t)) = 0$

$$
\varrho_r(x(t)) = \left(\frac{d_{left}(x(t)) + d_{right}(x(t))}{2d_{max}}\right)^2,
\tag{10.26}
$$

$$
\rho_r(X) = \sqrt{\frac{1}{T} \int_0^T \varrho_r(x(t))dt}, \quad P_r = \rho_r.
$$

where $d_{left} : \mathbb{R}^n \rightarrow \mathbb{R}^{\geq 0}$, $d_{right} : \mathbb{R}^n \rightarrow \mathbb{R}^{\geq 0}$ denote the left and right infringement distances of the ego AV footprint into the non-drivable areas, respectively, $\mathbb{R}^{\geq 0}$ denotes a non-negative real scalar. $d_{\max} > 0$ denotes the maximum infringement distance and is used to normalize the instantaneous violation score defined based on a L-2 norm, and ρ_r is the aggregation over trajectory duration T.

r_3 : Stay in lane

Statement: $d_{left}(\boldsymbol{x}(t)) + d_{right}(\boldsymbol{x}(t)) = 0$

$$\varrho_r(\boldsymbol{x}(t)) = \left(\frac{d_{left}(\boldsymbol{x}(t)) + d_{right}(\boldsymbol{x}(t))}{2d_{max}} \right)^2, \tag{10.27}$$

$$\rho_r(X) = \sqrt{\frac{1}{T} \int_0^T \varrho_r(\boldsymbol{x}(t))dt}, \quad P_r = \rho_r.$$

where $d_{left} : \mathbb{R}^n \rightarrow \mathbb{R}^{\geq 0}$, $d_{right} : \mathbb{R}^n \rightarrow \mathbb{R}^{\geq 0}$ denote the left and right infringement distances of the ego AV footprint into the left and right lane boundaries, respectively. The violation scores are defined similar to rule r_2.

r_4 : Satisfy the maximum speed limit

Statement: $v(t) \leq v_{max,s}$

$$\varrho_r(\boldsymbol{x}(t)) = \max\left(0, \frac{v(t) - v_{max,s}}{v_{max}} \right)^2, \tag{10.28}$$

$$\rho_r(X) = \sqrt{\frac{1}{T} \int_0^T \varrho_r(\boldsymbol{x}(t))dt}, \quad P_r = \rho_r.$$

where $v_{max,s} > 0$ denotes the maximum speed in a scenario s and varies for different road types (e.g., highway, residential, etc.).

r_5 : Satisfy the minimum speed limit

Statement: $v(t) \geq v_{min,s}$

$$\varrho_r(\boldsymbol{x}(t)) = \max(0, \frac{v_{min,s} - v(t)}{v_{min,s}})^2, \tag{10.29}$$

$$\rho_r(X) = \sqrt{\frac{1}{T} \int_0^T \varrho_r(\boldsymbol{x}(t))dt}, \quad P_r = \rho_r.$$

where $v_{min,s} > 0$ denotes the minimum speed in a scenario s, which is dependent on the road type.

r_6 : Drive smoothly

Statement: $|a(t)| \leq a_{max,s} \wedge |a_{lat}(t)| \leq a_{lat,s}$

$$\varrho_r(\boldsymbol{x}(t)) = \left(\max(0, \frac{a_{max,s} - |a(t)|}{a_{max}}) + \max(0, \frac{a_{lat,s} - |a_{lat}(t)|}{a_{lat_m}}) \right)^2 , \qquad (10.30)$$

$$\rho_r(\mathcal{X}) = \sqrt{\frac{1}{T} \int_0^T \varrho_r(\boldsymbol{x}(t))dt}, \quad P_r = \rho_r.$$

where $a_{lat}(t) = \kappa v^2(t)$ denotes the lateral acceleration at time instant t; $a_{max,s} > 0$, $a_{lat,s} > 0$ denote the maximum and the allowed lateral acceleration in a scenario s, respectively; and a_{max} and $a_{lat_m} > 0$ denote the maximum feasible acceleration and maximum feasible lateral acceleration of the vehicle, respectively.

r_7 : Maintain clearance with parked vehicles

Statement: $d_{min,fp}(\boldsymbol{x}, \boldsymbol{x}_i) \geq d_7 + v(t)\eta_7, \forall i \in S_{pveh}$

$$\varrho_{r,i}(\boldsymbol{x}(t)) = \max(0, \frac{d_7 + v(t)\eta_7 - d_{min,fp}(\boldsymbol{x}, \boldsymbol{x}_i)}{d_7 + v_{max}\eta_7})^2, \qquad (10.31)$$

$$\rho_{r,i}(\mathcal{X}) = \max_{t \in [0,T]} \varrho_i(\boldsymbol{x}(t)), \quad P_r = \sqrt{\frac{1}{n_{pveh}} \sum_{i \in S_{pveh}} \rho_{r,i}}$$

where $d_{min,fp} : \mathbb{R}^{n+n_i} \to \mathbb{R}$ denotes the distance between the footprints of the ego AV and the parked vehicle i, $d_7 \geq 0$, $\eta_7 > 0$, and violation scores are defined similar to r_1, S_{pveh} and $n_{pveh} \in \mathbb{N}$ denote the index set and number of parked vehicles, respectively, and $\boldsymbol{x}_i \in \mathbb{R}^{n_i}$ denotes the state of parked vehicle i.

r_8 : Maintain clearance with active vehicles

Statement: $\quad d_{min,l}(\boldsymbol{x}, \boldsymbol{x}_i) \geq d_{8,l} + v(t)\eta_{8,l}$

$\qquad \wedge \; d_{min,r}(\boldsymbol{x}, \boldsymbol{x}_i) \geq d_{8,r} + v(t)\eta_{8,r}$

$\qquad \wedge \; d_{min,f}(\boldsymbol{x}, \boldsymbol{x}_i) \geq d_{8,f} + v(t)\eta_{8,f}, \forall i \in S_{aveh}$

$$\varrho_{r,i}(\boldsymbol{x}(t)) = \frac{1}{3}(\max(0, \frac{d_{8,l} + v(t)\eta_{8,l} - d_{min,l}(\boldsymbol{x}, \boldsymbol{x}_i)}{d_{8,l} + v_{max}\eta_{8,l}})^2$$

$$+ \max(0, \frac{d_{8,r} + v(t)\eta_{8,r} - d_{min,r}(\boldsymbol{x}, \boldsymbol{x}_i)}{d_{8,r} + v_{max}\eta_{8,r}})^2 \qquad (10.32)$$

$$+ \max(0, \frac{d_{8,f} + v(t)\eta_{8,f} - d_{min,f}(\boldsymbol{x}, \boldsymbol{x}_i)}{d_{8,f} + v_{max}\eta_{8,f}})^2),$$

$$\rho_{r,i}(\mathcal{X}) = \frac{1}{T} \int_0^T \varrho_{r,i}(\boldsymbol{x}(t))dt, \quad P_r = \sqrt{\frac{1}{n_{aveh} - 1} \sum_{i \in S_{aveh} \setminus ego} \rho_{r,i}}$$

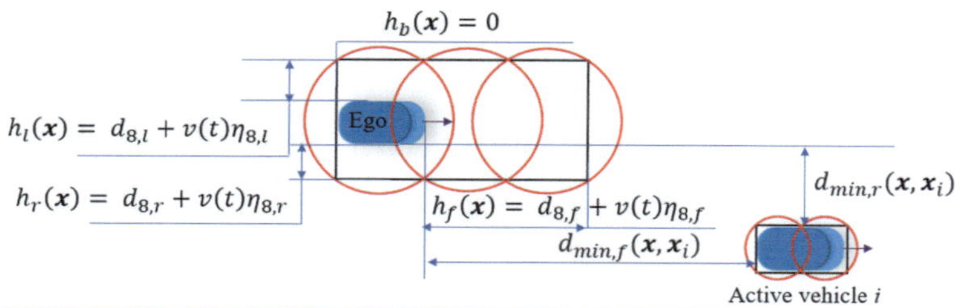

Fig. 10.7 Formulation of r_8 with the optimal disk coverage approach: r_8 is satisfied since the clearance regions of the ego AV and the active vehicle $i \in S_p$ do not overlap

where $d_{min,l} : \mathbb{R}^{n+n_i} \to \mathbb{R}, d_{min,r} : \mathbb{R}^{n+n_i} \to \mathbb{R}, d_{min,f} : \mathbb{R}^{n+n_i} \to \mathbb{R}$ denote the distance between the footprints of the ego AV and active vehicle i on the left, right and front, respectively; $d_{8,l} \geq 0, d_{8,r} \geq 0, d_{8,f} \geq 0, \eta_{8,l} > 0, \eta_{8,r} > 0, \eta_{8,f} > 0$ are defined similarly as in r_1. S_{aveh} and $n_{aveh} \in \mathbb{N}$ denote the index set and number of active vehicles, and $x_i \in \mathbb{R}^{n_i}$ denotes the state of active vehicle i. Similar to Fig. 10.4, we show in Fig. 10.7 how r_8 is defined based on the previous described clearance region and optimal disk coverage.

We consider the TORQ $\langle R, \sim_p, \leq_p \rangle$ from Fig. 10.8, with rules $R = \{r_1, r_2, r_3, r_4, r_5, r_6, r_7, r_8\}$. The optimal disk coverage is used to compute the optimal controls for all the clearance rules, which are implemented using HOCBFs.

In the following, we consider three common driving scenarios. For each of them, we solve the optimal control problem **OCP-R** and perform a pass/fail evaluation. In all three scenarios, in the pass/fail evaluation, an initial candidate trajectory is drawn "by hand" using the tool described below. We use CLFs to generate a feasible trajectory $\mathcal{X}_c$ which tracks the candidate trajectory subject to the vehicle dynamics (2.11), control bounds (2.19) and state constraints (10.7).

All the following examples were produced using a user-friendly software tool in MAT-LAB. The tool allows to load a map represented by a .json file and place vehicles and pedestrians on it. It provides an interface to generate smooth reference/candidate trajectories and it implements our proposed optimal control and pass/fail frameworks. The *quadprog* optimizer was used to solve the QPs (solve time < 0.01 s for each QP) and *ode*45 to integrate the vehicle dynamics (10.21). All computations were performed on a Intel(R) Core(TM) i7-8700 CPU @ 3.2 GHz×2. The simulation parameters used are as follows: $v_{max} = 10$ m/s, $a_{max} = -a_{min} = 3.5$ m/s^2, $u_{j,max} = -u_{j,min} = 4$ m/s^3, $\delta_{max} = -\delta_{min} = 1$ rad, $\omega_{max} = -\omega_{min} = 0.5$ rad/s, $u_{s,max} = -u_{s,min} = 2$ rad/s^2, $w = 1.8$ m, $l = 4$ m, $l_f = l_r = 2$ m, $d_1 = 1$ m, $\eta_1 = 0.067$ s, $v_{max,s} = 7$ m/s, $v_{min,s} = 3$ m/s, $a_{max,s} = 2.5$ m/s^2, $a_{lat_m} = 3.5$ m/s^2, $a_{lat,s} = 1.75$ m/s^2, $d_7 = 0.3$ m, $\eta_7 = 0.13$ s, $d_{8,l} = d_{8,r} = 0.5$ m, $d_{8,f} = 1$ m, $\eta_{8,r} = \eta_{8,l} = 0.036$ s, $\eta_{8,f} = 2$ s, $v_d = 4$ m/s, $\beta = 2$ in (10.14).

Fig. 10.8 TORQ priority structure for case study

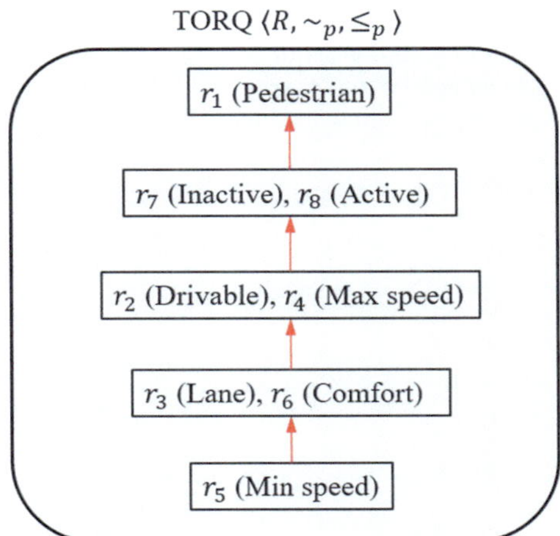

Scenario 1: Assume there is an active vehicle, a parked (inactive) vehicle and a pedestrian, as shown in Fig. 10.9.

Optimal control: We solve the optimal control problem (10.17) by starting the rule relaxation from $S_1 = \{\emptyset\}$ (i.e., without relaxing any rules). This problem is infeasible in the given scenario since the ego AV cannot maintain the required distance between both the active and the parked vehicles as the clearance rules are speed-dependent. Therefore, we relaxed the next lowest priority equivalence class set in S_{sorted}, i.e., the minimum speed limit rule in $S_2 = \{\{r_5\}\}$, for which it is possible to find a feasible trajectory as illustrated in Fig. 10.9. By checking δ_i for r_5 from (10.17), we found it is positive in some time intervals in $[0, T]$, thus, r_5 is indeed relaxed. The total violation score for rule r_5 from (10.29) for the generated trajectory is 0.539, and all other rules in R are satisfied. Thus, by Definition 10.4, the generated trajectory satisfies $\langle R, \sim_p, \leq_p \rangle$ in Fig. 10.8.

Pass/Fail: The candidate trajectory $\mathcal{X}_c$ is shown in Fig. 10.10. This candidate trajectory only violates rule r_5 with a total violation score 0.682. Following the proposed pass/fail approach, we can either relax r_5 or not relax any rules to find a possibly better trajectory. As shown in the above optimal control problem for this scenario, we cannot find a feasible solution if we do not relax rule r_5. However, since the violation of r_5 by the candidate trajectory is larger than that of the optimal trajectory in Fig. 10.9, we "fail" the candidate trajectory.

Scenario 2: Assume there is an active vehicle, two parked (inactive) vehicles and two pedestrians, as shown in Fig. 10.11.

Optimal control: Similar to Scenario 1, the optimal control problem (10.17) starting from $S_1 = \{\emptyset\}$ (without relaxing any rules in R) is infeasible. We relax the next lowest priority rule set in S_{sorted}, i.e., the minimum speed rule in $S_2 = \{\{r_5\}\}$, for which it is possible to find a feasible trajectory as illustrated in Fig. 10.11. Again, the value of δ_i for r_5 is positive

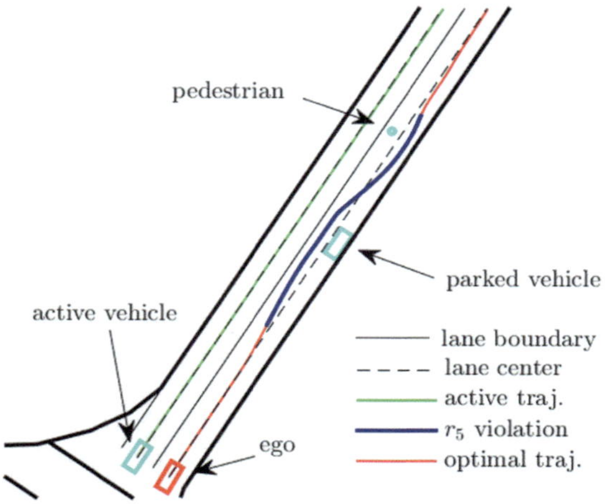

Fig. 10.9 Optimal control for Scenario 1: the subset of optimal ego AV trajectories violating r_5 is shown in blue

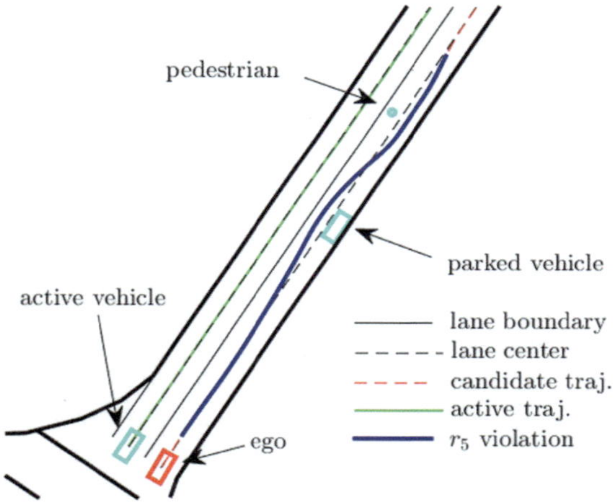

Fig. 10.10 Pass/Fail for Scenario 1: the subset of candidate trajectories violating r_5 is shown in blue; the alternative trajectory in this scenario is the same as in Fig. 10.9

in some time intervals in $[0, T]$, thus, r_5 is indeed relaxed. The total violation score of the rule r_5 for the generated trajectory is 0.646, and all other rules in R are satisfied.

Pass/Fail: The candidate trajectory $\mathcal{X}_c$ shown as a red dashed line in Fig. 10.12 (left) violates rules r_1, r_3 and r_8 with total violation scores 0.01, 0.23, 0.22 found from (10.25), (10.27), (10.32), respectively. In this scenario, we know that the ego AV can change lanes (where the lane keeping rule r_3 is in a lower priority equivalence class than r_1) to get a reasonable trajectory. Thus, we show the case of relaxing the rules in the equivalence classes $O_2 = \{r_3, r_6\}$ and $O_1 = \{r_5\}$ to find a feasible trajectory that is better than the candidate one. The optimal control problem (10.17) generates a trajectory as the red-solid curve shown in Fig. 10.12, and only δ_i for r_6 is 0 for all $[0, T]$. Thus, r_6 does not need to be relaxed. The generated trajectory violates rules r_3 and r_5 with total violation scores 0.124 and 0.111,

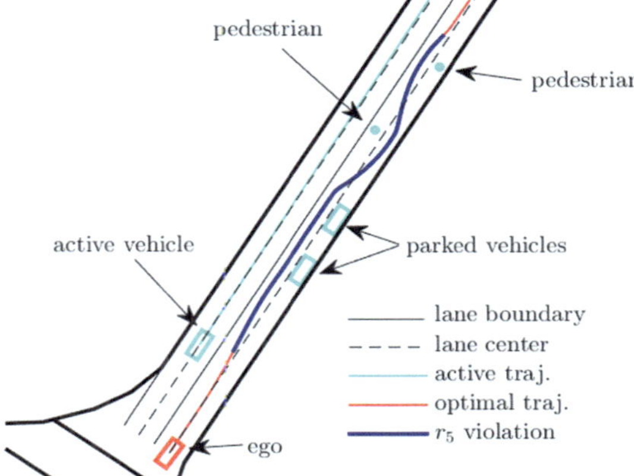

Fig. 10.11 Optimal control for Scenario 2: the subset of optimal ego trajectories violating r_5 is shown in blue

respectively, but satisfies all the other rules including the highest priority rule r_1. According to Definition 10.3 for the given $\langle R, \sim_p, \leq_p \rangle$ in Fig. 10.8, the new generated trajectory is better than the candidate one, thus, we fail the candidate trajectory. Note that although this trajectory violates the lane keeping rule, it has a smaller violation score for r_5 compared to the trajectory obtained from the optimal control in Fig. 10.11 (0.111 versus 0.646), i.e., the average speed of ego in the red-solid trajectory in Fig. 10.12 is larger.

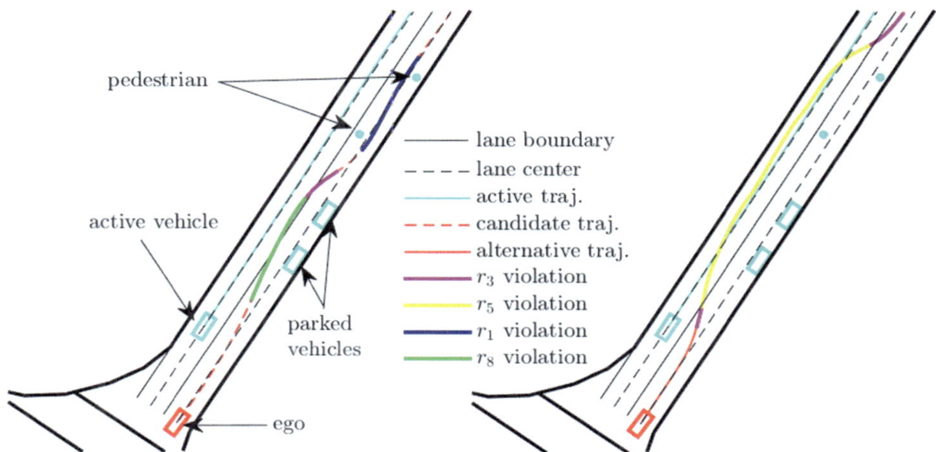

Fig. 10.12 Pass/Fail for Scenario 2: the subsets of the candidate trajectory (left) violating r_8, r_3, r_1 are shown in green, magenta and blue, respectively; the subsets of alternative trajectory (right) violating r_5, r_3 are shown in yellow and magenta, respectively

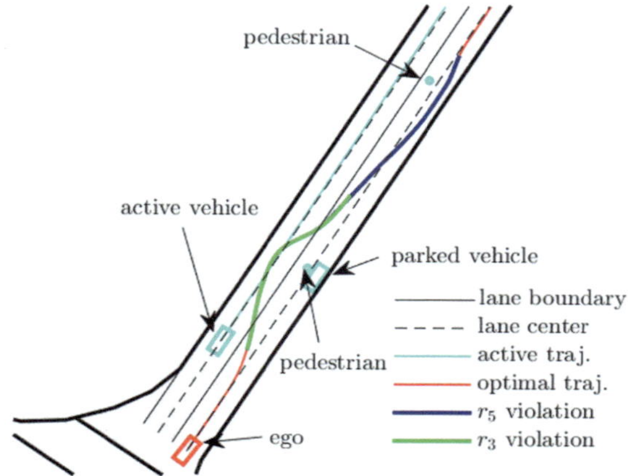

Fig. 10.13 Optimal control for Scenario 3: the subsets of the optimal ego AV trajectory violating r_5, r_3 are shown in blue and green, respectively

Scenario 3: Assume there is an active vehicle, a parked vehicle and two pedestrians (one just gets out of the parked vehicle), as shown in Fig. 10.13.

Optimal control: Similar to Scenario 1, the optimal control problem (10.17) starting from $S_1 = \{\emptyset\}$ (without relaxing any rules in R) is infeasible. We relax the lowest priority rule set in S_{sorted}, i.e., the minimum speed rule $S_2 = \{\{r_5\}\}$, and solve the optimal control problem. In the (feasible) generated trajectory, the ego AV stops before the parked vehicle, which satisfies all the rules in R except r_5. Thus, by Definition 10.4, the generated trajectory satisfies the TORQ $\langle R, \sim_p, \leq_p \rangle$. However, this might not be a desirable behavior, thus, we further relax the lane keeping r_3 and comfort r_6 rules and find the feasible trajectory shown in Fig. 10.13. In this case, δ_i for r_6 is 0 for all $[0, T]$, therefore, r_6 does not need to be relaxed. The total violation scores for the rules r_3 and r_5 are 0.058 and 0.359, respectively, and all other rules in R are satisfied.

Pass/Fail: The candidate trajectory $\mathcal{X}_c$ shown as the red-dashed curve in Fig. 10.14 violates rules r_3 and r_8 with total violation scores 0.025 and 0.01, respectively. In this scenario, from the optimal control in Fig. 10.13 we know that ego can change lane (where the lane keeping rule is in a lower priority equivalence class than r_8). We show the case of relaxing the rules in the equivalence classes $O_2 = \{r_3, r_6\}$ and $O_1 = \{r_5\}$ (all have lower priorities than r_8). The optimal control problem (10.17) generates the red-solid curve shown in Fig. 10.14. By checking δ_i for r_6, we find that r_6 is indeed not relaxed. The generated alternative trajectory violates rules r_3 and r_5 with total violation scores 0.028 and 0.742, respectively, but satisfies all other rules including r_8. According to Definition 10.3 and Fig. 10.8, the new generated trajectory (although violates r_3 more than the candidate trajectory, it does not violate r_8 which has a higher priority) is better than the candidate one. Thus, we fail the candidate trajectory.

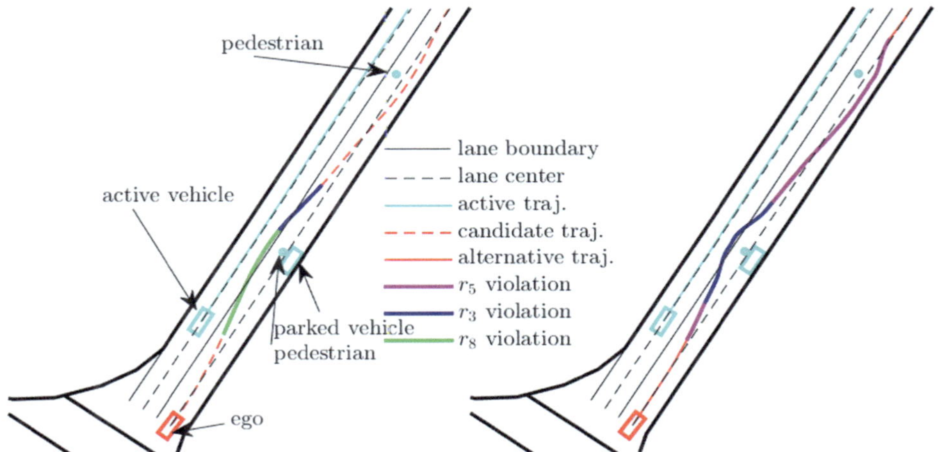

Fig. 10.14 Pass/Fail for Scenario 3: the subsets of the candidate trajectory (left) violating r_8, r_3 are shown in green and blue, respectively; the subsets of the alternative trajectory (right) violating r_5, r_3 are shown in magenta and blue, respectively

10.3 Quadrotor Safe Navigation

In this section, we consider a quadrotor modeled through a set of standard equations capturing its dynamics and employ HOCBFs to guarantee safe navigation.

The quadrotor dynamics are defined in (10.33) through (10.35), capturing the position dynamics, attitude dynamics [36, 61], and rotor dynamics [11], respectively. The inertial frame of the quadrotor denotes the global/earth frame, while the body frame denotes the frame attached to the quadrotor itself.

$$
\begin{bmatrix} \dot{x} \\ \dot{v}_x \\ \dot{y} \\ \dot{v}_y \\ \dot{z} \\ \dot{v}_z \end{bmatrix} = \begin{bmatrix} v_x \\ 0 \\ v_y \\ 0 \\ v_z \\ -g \end{bmatrix} + \frac{F}{M} \begin{bmatrix} 0 \\ C_\psi S_\theta C_\phi + S_\psi S_\phi \\ 0 \\ S_\psi S_\theta C_\phi - C_\psi S_\phi \\ 0 \\ C_\theta C_\phi \end{bmatrix} - \frac{1}{M} \begin{bmatrix} 0 & 0 & 0 \\ A_x & 0 & 0 \\ 0 & 0 & 0 \\ 0 & A_y & 0 \\ 0 & 0 & 0 \\ 0 & 0 & A_z \end{bmatrix} \begin{bmatrix} v_x \\ v_y \\ v_z \end{bmatrix}, \qquad (10.33)
$$

$$
\begin{bmatrix} \dot{\phi} \\ \dot{\omega}_\phi \\ \dot{\theta} \\ \dot{\omega}_\theta \\ \dot{\psi} \\ \dot{\omega}_\psi \end{bmatrix} = \begin{bmatrix} \omega_\phi & 0 & 0 \\ 0 & \omega_\phi C_\phi T_\theta + \omega_\theta S_\phi/C_\theta^2 & -\omega_\phi S_\phi C_\theta + \omega_\theta C_\phi/C_\theta^2 \\ \omega_\theta & 0 & 0 \\ 0 & -\omega_\phi S_\phi & -\omega_\phi C_\phi \\ \omega_\psi & 0 & 0 \\ 0 & \omega_\phi C_\phi/C_\theta + \omega_\theta S_\phi T_\theta/C_\theta & -\omega_\phi S_\phi/C_\theta + \omega_\theta C_\phi T_\theta/C_\theta \end{bmatrix} \begin{bmatrix} 1 \\ q \\ r \end{bmatrix} + W_\eta^{-1} \begin{bmatrix} \dot{p} \\ \dot{q} \\ \dot{r} \end{bmatrix},
$$

$$
(10.34)
$$

$$\dot{\omega}_i = b_1 u_i - \beta_0 - \beta_1 \omega_i - \beta_2 \omega_i^2, \, i \in \{1, 2, 3, 4\}. \tag{10.35}$$

In (10.33), $(x, y, z) \in \mathbb{R}^3$ denotes the 3-D location of the quadrotor in the inertial frame, $(v_x, v_y, v_z) \in \mathbb{R}^3$ denotes the speed vector along (x, y, z) in the inertial frame, and $A_x > 0, A_y > 0, A_z > 0$ are the drag force coefficients for velocities in the corresponding directions of the inertial frame. Moreover, M denotes the mass of the quadrotor, g is the gravity constant. Note that C_{angle}, S_{angle} are abbreviations for $\cos(angle), \sin(angle)$, respectively. The angles $\phi \in \mathbb{R}, \theta \in \mathbb{R}, \psi \in \mathbb{R}$ denote the roll, pitch, and yaw angles (Euler angles) of the quadrotor in the inertial frame, respectively. Finally, F denotes the thrust from the four rotors in the direction of the $z-$axis in the body frame of the quadrotor, and is defined as

$$F = k \sum_{i=1}^{4} \omega_i^2, \tag{10.36}$$

where $k > 0$ denotes the thrust constant, and ω_i denotes the angular speed of rotor $i \in \{1, 2, 3, 4\}$.

In (10.34), $\omega_\phi \in \mathbb{R}, \omega_\theta \in \mathbb{R}, \omega_\psi \in \mathbb{R}$ denote the angular speeds of the quadrotor along ϕ, θ, ψ in the inertial frame, respectively. $(p, q, r) \in \mathbb{R}^3$ denotes the speed vector of the quadrotor along roll, pitch, yaw in the body frame. T_θ is short for $\tan \theta$. W_η^{-1} is a matrix defined as

$$W_\eta^{-1} = \begin{bmatrix} 1 & S_\phi T_\theta & C_\phi T_\theta \\ 0 & C_\phi & -S_\theta \\ 0 & S_\phi/C_\theta & C_\phi/C_\theta \end{bmatrix}, \tag{10.37}$$

The vector $[\dot{p}, \dot{q}, \dot{r}]^T$ in (10.34) is defined as

$$\begin{bmatrix} \dot{p} \\ \dot{q} \\ \dot{r} \end{bmatrix} = \begin{bmatrix} (I_{yy} - I_{zz})qr/I_{xx} \\ (I_{zz} - I_{xx})pr/I_{yy} \\ (I_{xx} - I_{yy})pq/I_{zz} \end{bmatrix} - I_r \begin{bmatrix} q/I_{xx} \\ -p/I_{yy} \\ 0 \end{bmatrix} \omega_\Gamma + \begin{bmatrix} \tau_\phi/I_{xx} \\ \tau_\theta/I_{yy} \\ \tau_\psi/I_{zz} \end{bmatrix}, \tag{10.38}$$

where $\omega_\Gamma = \omega_1 - \omega_2 + \omega_3 - \omega_4$, I_{xx}, I_{yy}, I_{zz} denote the inertial moment of the quadrotor along x, y, z axis respectively. I_r denotes the inertial moment of the rotor. $\tau_\phi, \tau_\theta, \tau_\psi$ denote the torques in the direction of the corresponding body frame angles, and are given by

$$\begin{bmatrix} \tau_\phi \\ \tau_\theta \\ \tau_\psi \end{bmatrix} = \begin{bmatrix} lk(-\omega_2^2 + \omega_4^2) \\ lk(-\omega_1^2 + \omega_3^2) \\ b_0(-\omega_1^2 + \omega_2^2 - \omega_3^2 + \omega_4^2) \end{bmatrix}, \tag{10.39}$$

where $b_0 > 0$ denotes the drag constant, and $l > 0$ is the distance between the rotor and the center of mass of the quadrotor.

In (10.35), $b_1 = \frac{k_m}{RI_r}, \beta_0 = \frac{C_s}{I_r}, \beta_1 = \frac{k_e k_m}{RI_r}, \beta_2 = \frac{k_r}{I_r}$, where k_e, k_m denote the electrical and mechanical torque constants, respectively, k_r denotes the load constant torque, C_s denotes the solid friction, and R denotes the motor internal resistance. u_i denotes the input (voltage) of the rotor $i \in \{1, 2, 3, 4\}$ (control inputs of the quadrotor).

In the safe navigation problem we consider, there are two **objectives**: (i) *Minimize energy consumption*

$$J(\boldsymbol{u}(t)) = \int_{t_0}^{t_f} [u_1^2(t) + u_2^2(t) + u_3^2(t) + u_4^2(t)]dt,$$

and (ii) *Reach destination* $(x_d, y_d, z_d) \in \mathbb{R}^3$, over a time interval $[t_1, t_2]$, $t_0 \le t_1 \le t_2 \le t_f$.
There are also two types of **constraints**: (i) *Safety* expressed as

$$(x(t) - x_o)^2 + (y(t) - y_o)^2 + (z(t) - y_o)^2 \ge r_0^2, \tag{10.40}$$

where $(x_o, y_o, z_o) \in \mathbb{R}^3$ denotes the location of a spherical obstacle and r_0 is its size, and (ii) *Quadrotor limitations*, consisting of the minimum and maximum speed limits on v_x, v_y, v_z, the minimum and maximum angular limits on pitch and roll, respectively.

Our goal is to determine control laws to achieve the two objectives above subject to the safety constraint (10.40) and the quadrotor limitations (both must be strictly satisfied), for the quadrotor governed by dynamics (10.33) through (10.35).

Based on the current location of the quadrotor and its desired destination, we can specify desired angles for roll and pitch, and a desired height for the state variable z. Then, we define three CLFs to stabilize the above three states to the desired values. In addition, we also define a CLF to stabilize the yaw rotation speed r to zero. We define HOCBFs to enforce the quadrotor limitations.

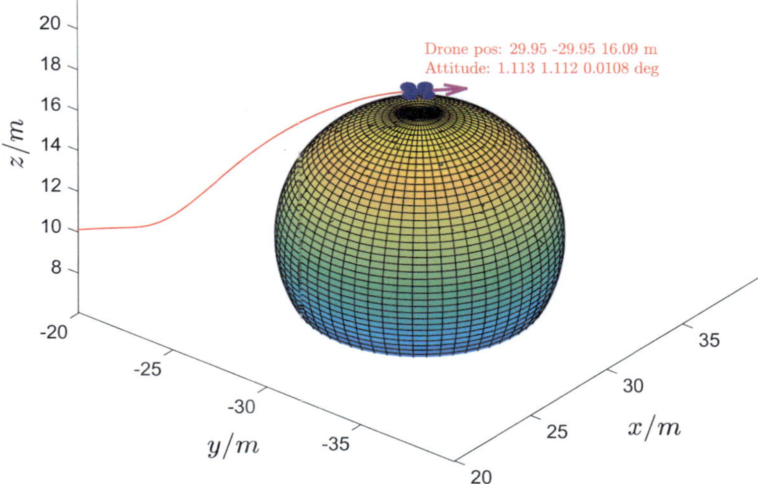

Fig. 10.15 Quadrotor (drone) trajectory under the HOCBF method with a spherical obstacle defined as in (10.40)

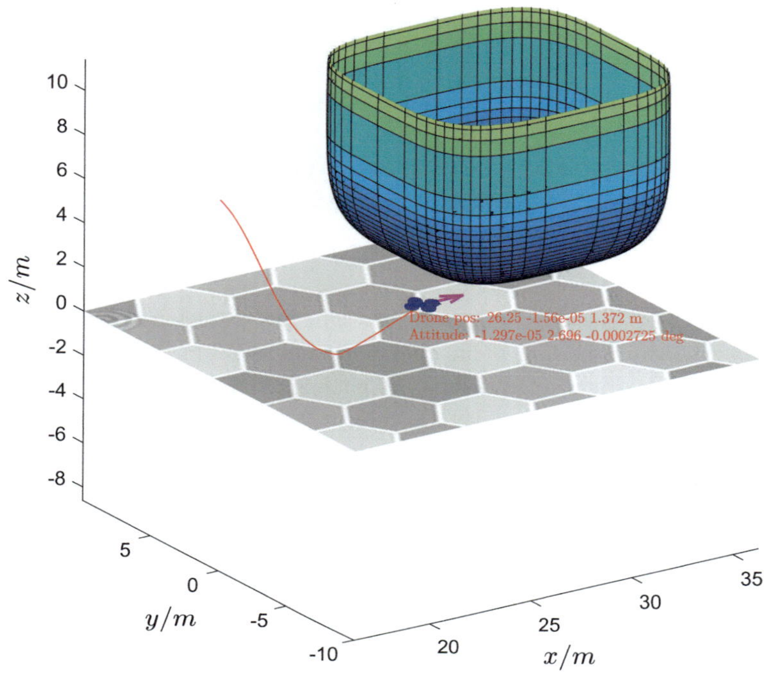

Fig. 10.16 Quadrotor (drone) trajectory under the HOCBF method with a superellipse-type obstacle. The safety constraint (10.40) in this case becomes $(x(t) - x_o)^4 + (y(t) - y_o)^4 + (z(t) - y_o)^4 \geq r_0^4$

As for the safety constraint (10.40), we use a HOCBF to enforce it. The relative degree of (10.40) is three. The corresponding HOCBF constraint is too complicated to provide an explicit expression here due to the complex dynamics (10.33)–(10.35), thus we omit it.

The simulation parameters used are: $\boldsymbol{x}(0) = (x(0), v_x(0), y(0), v_y(0), z(0), v_z(0), \phi(0), \omega_\phi(0), \theta(0), \omega_\theta(0), \psi(0), \omega_\psi(0), \omega_1(0), \omega_2(0), \omega_3(0), \omega_4(0)) = (0, 0, 0, 0, 0.01, 0, 0, 0, 0, 0, 0, 0, 622.2, 622.2, 622.2, 622.2)$, $(x_d, y_d, z_d) = (50, -50, 10)$ m, $(x_o, y_o, z_o) = (30, -30, 10)$ m, $r_0 = 7$ m, $M = 0.468$ kg, $g = 9.81$ ms/s^2, $I_{xx} = I_{yy} = 4.856\,e^{-3}$kg $\cdot$ m^2, $I_{zz} = 8.801\,e^{-3}$kg $\cdot$ m^2, $I_r = 3.357\,e^{-5}$kg $\cdot$ m^2, $l = 0.225$ m, $k = 2.980\,e^{-6}$, $b_0 = 1.140\,e^{-7}$, $b_1 = 280.19$, $\beta_0 = 189.63$, $\beta_1 = 6.0612$, $\beta_2 = 0.0122$, $k_1 = 0.5$, $p_1 = 5\,e^{-7}$, $p^* = 0.5$.

The simulation trajectories of the quadrotor are shown in Fig. 10.15. As expected, the HOCBF method can guarantee safety (collision-free trajectories) for the quadrotor. We have also tested a superellipse-type (cube-like) obstacle with the associated safe simulated trajectory shown in Fig. 10.16.

References

1. Ames, A.D., Galloway, K., Grizzle, J.W.: Control Lyapunov functions and hybrid zero dynamics. In: Proceedings of the 51rd IEEE Conference on Decision and Control, pp. 6837–6842 (2012)
2. Ames, A.D., Grizzle, J.W., Tabuada, P.: Control barrier function based quadratic programs with application to adaptive cruise control. In: Proceedings of the 53rd IEEE Conference on Decision and Control, pp. 6271–6278 (2014)
3. Ames, A.D., Notomista, G., Wardi, Y., Egerstedt, M.: Integral control barrier functions for dynamically defined control laws. IEEE Control Syst. Lett. **5**(3), 887–892 (2020)
4. Ames, A.D., Xu, X., Grizzle, J.W., Tabuada, P.: Control barrier function based quadratic programs for safety critical systems. IEEE Trans. Autom. Control **62**(8), 3861–3876 (2017)
5. Athans, M.: A unified approach to the vehicle-merging problem. Transp. Res. **3**(1), 123–133 (1969)
6. Baier, C., Katoen, J.: Principles of Model Checking. The MIT Press (2008)
7. Belta, C., Sadraddini, S.: Formal methods for control synthesis: an optimization perspective. Annu. Rev. Control, Robot., Auton. Syst. **2**(1), 115–140 (2019). https://doi.org/10.1146/annurev-control-053018-023717
8. Belta, C., Yordanov, B., Gol, E.A.: Formal Methods for Discrete-Time Dynamical Systems. Springer, Berlin (2017)
9. Bertsimas, D., Tsitsiklis, J.N.: Introduction to Linear Optimization, vol. 6. Athena Scientific Belmont, MA (1997)
10. Bishop, C.M.: Pattern Recognition and Machine Learning. Springer, Berlin (2006)
11. Bouadi, H., Bouchoucha, M., Tadjine, M.: Sliding mode control based on backstepping approach for an UAV type-quadrotor. World Acad. Sci., Eng. Technol. **26**(5), 22–27 (2007)
12. Bryson, H.: Applied Optimal Control. Ginn Blaisdell, Waltham, MA (1969)
13. Censi, A., Slutsky, K., Wongpiromsarn, T., Yershov, D., Pendleton, S., Fu, J., Frazzoli, E.: Liability, ethics, and culture-aware behavior specification using rulebooks. In: 2019 International Conference on Robotics and Automation, pp. 8536–8542 (2019)

W. Xiao et al., *Safe Autonomy with Control Barrier Functions*, Synthesis Lectures on Computer Science, https://doi.org/10.1007/978-3-031-27576-0

14. Chalaki, B., Beaver, L.E., Malikopoulos, A.A.: Experimental validation of a real-time optimal controller for coordination of CAVs in a multi-lane roundabout. In: 2020 IEEE Intelligent Vehicles Symposium (IV), pp. 775–780. IEEE (2020)

15. Chen, L., Englund, C.: Cooperative intersection management: a survey. IEEE Trans. Intell. Transp. Syst. **17**(2), 570–586 (2015)

16. Clarke, E.M., Grumberg, O., Peled, D.A.: Model Checking. MIT Press (1999)

17. Collin, A., Bilka, A., Pendleton, S., Tebbens, R.D.: Safety of the intended driving behavior using rulebooks. In: IV Workshop on Ensuring and Validating Safety for Automated Vehicles, pp. 1–7 (2020)

18. Donzé, A., Maler, O.: Robust satisfaction of temporal logic over real-valued signals. In: International Conference on Formal Modeling and Analysis of Timed Systems, pp. 92–106 (2010)

19. Fan, D.D., Nguyen, J., Thakker, R., Alatur, N., Agha-Mohammadi, A.A., Theodorou, E.A.: Bayesian learning-based adaptive control for safety critical systems. In: 2020 IEEE International Conference on Robotics and Automation, pp. 4093–4099 (2020)

20. Flannery, A., Datta, T.: Operational performance measures of American roundabouts. Transp. Res. Rec. **1572**(1), 68–75 (1997)

21. Glotfelter, P., Cortes, J., Egerstedt, M.: Nonsmooth barrier functions with applications to multi-robot systems. IEEE Control Syst. Lett. **1**(2), 310–315 (2017)

22. Hart, P.E., Nilsson, N.J., Raphael, B.: A formal basis for the heuristic determination of minimum cost paths. IEEE Trans. Syst. Sci. Cybern. **4**(2), 100–107 (1968)

23. ISO: Pas 21448-road vehicles-safety of the intended functionality. International Organization for Standardization (2019)

24. Kamal, M., Mukai, M., Murata, J., Kawabe, T.: Model predictive control of vehicles on urban roads for improved fuel economy. IEEE Trans. Control Syst. Technol. **21**(3), 831–841 (2013)

25. Kaufman, H., Barkana, I., Sobel, K.: Direct adaptive control algorithms: theory and applications. Springer Science & Business Media (2012)

26. Khalil, H.K.: Nonlinear Systems, 3rd edn. Prentice Hall (2002)

27. Khojasteh, M.J., Dhiman, V., Franceschetti, M., Atanasov, N.: Probabilistic safety constraints for learned high relative degree system dynamics. In: Learning for Dynamics and Control, pp. 781–792 (2020)

28. Köhler, J., Kötting, P., Soloperto, R., Allgöwer, F., Müller, M.A.: A robust adaptive model predictive control framework for nonlinear uncertain systems. Int. J. Robust Nonlinear Control **31**(18) (2021)

29. LaValle, S.M., Kuffner, J., James, J.: Randomized kinodynamic planning. Int. J. Robot. Res. **20**(5), 378–400 (2001)

30. Lindemann, L., Dimarogonas, D.V.: Control barrier functions for multi-agent systems under conflicting local signal temporal logic tasks. IEEE Control Syst. Lett. **3**(3), 757–762 (2019)

31. Lindemann, L., Dimarogonas, D.V.: Control barrier functions for signal temporal logic tasks. IEEE Control Syst. Lett. **3**(1), 96–101 (2019)

32. Lopez, B.T.: Adaptive robust model predictive control for nonlinear systems. Ph.D. thesis, Massachusetts Institute of Technology (2019)

33. Lopez, B.T., Slotine, J.J.E., How, J.P.: Robust adaptive control barrier functions: an adaptive and data-driven approach to safety. IEEE Control Syst. Lett. **5**(3), 1031–1036 (2020)

34. Lopez, P.A., Behrisch, M., Bieker-Walz, L., Erdmann, J., Flötteröd, Y.P., Hilbrich, R., Lücken, L., Rummel, J., Wagner, P., Wießner, E.: Microscopic traffic simulation using SUMO. In: 2018 21st International Conference on Intelligent Transportation Systems (ITSC), pp. 2575–2582. IEEE (2018)

35. Lu, X., Cannon, M., Koksal-Rivet, D.: Robust adaptive model predictive control: performance and parameter estimation. Int. J. Robust Nonlinear Control **31**(18), 8703–8724 (2021)

36. Luukkonen, T.: Modelling and control of quadcopter. In: Independent Research Project in Applied Mathematics, Espoo, vol. 22, p. 22 (2011)
37. Maler, O., Nickovic, D.: Monitoring temporal properties of continuous signals. In: Proceedings of the International Conference on FORMATS-FTRTFT, pp. 152–166 (2004)
38. Martin-Gasulla, M., García, A., Moreno, A.T.: Benefits of metering signals at roundabouts with unbalanced flow: patterns in Spain. Transp. Res. Rec. **2585**(1), 20–28 (2016)
39. Mehdipour, N., Vasile, C., Belta, C.: Arithmetic-geometric mean robustness for control from signal temporal logic specifications. In: American Control Conference, pp. 1690–1695 (2019)
40. Mitra, S.: Verifying Cyber-Physical Systems: A Path to Safe Autonomy. MIT Press (2021)
41. Mukai, M., Natori, H., Fujita, M.: Model predictive control with a mixed integer programming for merging path generation on motor way. In: Proceedings of the IEEE Conference on Control Technology and Applications (2017)
42. Nagumo, M.: Über die lage der integralkurven gewöhnlicher differentialgleichungen. In: Proceedings of the Physico-Mathematical Society of Japan, 3rd Series, vol. 24, pp. 551–559 (1942)
43. Nguyen, Q., Sreenath, K.: Exponential control barrier functions for enforcing high relative-degree safety-critical constraints. In: Proceedings of the American Control Conference, pp. 322–328 (2016)
44. Nocedal, J., Wright, S.J.: Numerical Optimization. Springer, Berlin (1999)
45. Nolte, M., Bagschik, G., Jatzkowski, I., Stolte, T., Reschka, A., Maurer, M.: Towards a skill-and ability-based development process for self-aware automated road vehicles. In: 2017 IEEE 20th International Conference on Intelligent Transportation Systems, pp. 1–6 (2017)
46. Nor, M.H.B.M., Namerikawa, T.: Merging of connected and automated vehicles at roundabout using model predictive control. In: Annual Conference of the Society of Instrument and Control Engineers of Japan, pp. 272–277 (2018)
47. Parseh, M., Asplund, F., Nybacka, M., Svensson, L., Törngren, M.: Pre-crash vehicle control and manoeuvre planning: a step towards minimizing collision severity for highly automated vehicles. In: 2019 IEEE International Conference of Vehicular Electronics and Safety (ICVES), pp. 1–6 (2019)
48. Prajna, S., Jadbabaie, A.: Safety verification of hybrid systems using barrier certificates. In Hybrid Systems: Computation and Control, pp. 477–492 (2004)
49. Prajna, S., Jadbabaie, A., Pappas, G.J.: A framework for worst-case and stochastic safety verification using barrier certificates. IEEE Trans. Autom. Control **52**(8), 1415–1428 (2007)
50. Qian, X., Gregoire, J., Moutarde, F., Fortelle, A.D.L.: Priority-based coordination of autonomous and legacy vehicles at intersection. In: IEEE Conference on Intelligent Transportation Systems, pp. 1166–1171 (2014)
51. Rawlings, J.B., Mayne, D.Q., Diehl, M.M.: Model Predictive Control: Theory, Computation, and Design, 2nd edn. Nob Hill Publishing (2017)
52. Rios-Torres, J., Malikopoulos, A.A.: A survey on the coordination of connected and automated vehicles at intersections and merging at highway on-ramps. IEEE Trans. Intell. Transp. Syst. **18**(5), 1066–1077 (2016)
53. Rucco, A., Notarstefano, G., Hauser, J.: An efficient minimum-time trajectory generation strategy for two-track car vehicles. IEEE Trans. Control Syst. Technol. **23**(4), 1505–1519 (2015)
54. Sardain, P., Bessonnet, G.: Forces acting on a biped robot. center of pressure-zero moment point. IEEE Trans. Syst., Man, Cybern.-Part A: Syst. Hum. **34**(5), 630–637 (2004)
55. Schrank, B., Eisele, B., Lomax, T., Bak, J.: The 2015 urban mobility scorecard. Texas A&M Transportation Institute (2015). http://mobility.tamu.edu
56. Shalev-Shwartz, S., Shammah, S., Shashua, A.: On a formal model of safe and scalable self-driving cars (2017). arXiv:1708.06374

57. Srinivasan, M., Coogan, S., Egerstedt, M.: Control of multi-agent systems with finite time control barrier certificates and temporal logic. In: 2018 IEEE Conference on Decision and Control (CDC), pp. 1991–1996. IEEE (2018)

58. Tabuada, P.: Event-triggered real-time scheduling of stabilizing control tasks. IEEE Trans. Autom. Control **52**(9), 1680–1685 (2007)

59. Tabuada, P.: Verification and Control of Hybrid Systems-A Symbolic Approach. Springer, Berlin (2009)

60. Tanaskovic, M., Fagiano, L., Smith, R., Morari, M.: Adaptive receding horizon control for constrained MIMO systems. Automatica **50**, 3019–3029 (2014)

61. Tayebi, A., McGilvray, S.: Attitude stabilization of a four-rotor aerial robot. In: 2004 43rd IEEE Conference on Decision and Control, vol. 2, pp. 1216–1221 (2004)

62. Taylor, A.J., Ames, A.D.: Adaptive safety with control barrier functions. In: Proceedings of the American Control Conference, pp. 1399–1405 (2020)

63. Taylor, A.J., Ong, P., Cortes, J., Ames, A.D.: Safety-critical event triggered control via input-to-state safe barrier functions. IEEE Control Syst. Lett. **5**(3), 749–754 (2021)

64. Tideman, M., van der Voort, M., van Arem, B., Tillema, F.: A review of lateral driver support systems. In: Proceedings of the IEEE Intelligent Transportation Systems Conference, pp. 992–999, Seatle (2007)

65. Ulbrich, S., Maurer, M.: Probabilistic online POMDP decision making for lane changes in fully automated driving. In: IEEE Conference on Intelligent Transportation Systems, pp. 2063–2067 (2013)

66. Vaidya, P.M.: Speeding-up linear programming using fast matrix multiplication. In: 30th Annual Symposium on Foundations of Computer Science, pp. 332–337 (1989)

67. Vogel, K.: A comparison of headway and time to collision as safety indicators. Accid. Anal. Prev. **35**(3), 427–433 (2003)

68. Waard, D.D., Dijksterhuis, C., Broohuis, K.A.: Merging into heavy motorway traffic by young and elderly drivers. Accid. Anal. Prev. **41**(3), 588–597 (2009)

69. Wiedemann, R.: Simulation des straßenverkehrsflusses. In: Proceedings of the Schriftenreihe des tnstituts fir Verkehrswesen der Universitiit Karlsruhe (In German language) (1974)

70. Wisniewski, R., Sloth, C.: Converse barrier certificate theorem. In: Proceedings of 52nd IEEE Conference on Decision and Control, pp. 4713–4718. Florence, Italy (2013)

71. Xiao, W., Belta, C.: Control barrier functions for systems with high relative degree. In: Proceedings of 58th IEEE Conference on Decision and Control, pp. 474–479. Nice, France (2019)

72. Xiao, W., Belta, C.: High order control barrier functions. IEEE Trans. Autom. Control **67**(7), 3655–3662 (2022)

73. Xiao, W., Belta, C., Cassandras, C.G.: Decentralized merging control in traffic networks: a control barrier function approach. In: Proceedings of the ACM/IEEE International Conference on Cyber-Physical Systems, pp. 270–279 (2019)

74. Xiao, W., Belta, C., Cassandras, C.G.: Feasibility guided learning for constrained optimal control problems. In: Proc. of 59th IEEE Conference on Decision and Control, pp. 1896–1901 (2020)

75. Xiao, W., Belta, C., Cassandras, C.G.: Event-triggered safety-critical control for systems with unknown dynamics. In: 2021 60th IEEE Conference on Decision and Control, pp. 540–545. IEEE (2021)

76. Xiao, W., Belta, C., Cassandras, C.G.: High order control Lyapunov-barrier functions for temporal logic specifications. In: Proceedings of the American Control Conference, pp. 4886–4891 (2021)

77. Xiao, W., Belta, C., Cassandras, C.G.: Adaptive control barrier functions. IEEE Trans. Autom. Control **67**(5), 2267–2281 (2022)

78. Xiao, W., Belta, C., Cassandras, C.G.: Event-triggered control for safety-critical systems with unknown dynamics. IEEE Trans. Autom. Control 1–16 (2022). https://doi.org/10.1109/TAC.2022.3202088

79. Xiao, W., Belta, C., Cassandras, C.G.: Sufficient conditions for feasibility of optimal control problems using control barrier functions. Automatica **135**, 109960 (2022)

80. Xiao, W., C. Belta, C.G.C., Rus, D.: Control barrier functions for systems with multiple control inputs. In: Proceedings of the American Control Conference (2022)

81. Xiao, W., Cassandras, C.G.: Conditions for improving the computational efficiency of decentralized optimal merging controllers for connected and automated vehicles. In: Proceedings of the 58th IEEE Conference on Decision and Control, pp. 3158–3163. Nice, France (2019)

82. Xiao, W., Cassandras, C.G.: Decentralized optimal merging control for connected and automated vehicles with optimal dynamic resequencing. In: Proceedings of the American Control Conference, pp. 4090–4095 (2020)

83. Xiao, W., Cassandras, C.G.: Decentralized optimal merging control for connected and automated vehicles with safety constraint guarantees. Automatica **123**, 109333 (2021)

84. Xiao, W., Cassandras, C.G., Belta, C.: Decentralized merging control in traffic networks with noisy vehicle dynamics: a joint optimal control and barrier function approach. In: Proceedings of the IEEE 22nd Intelligent Transportation Systems Conference (2019)

85. Xiao, W., Cassandras, C.G., Belta, C.: Decentralized optimal control in multi-lane merging for connected and automated vehicles. In: Proceedings of the IEEE 23nd Intelligent Transportation Systems Conference, pp. 1–6 (2020)

86. Xiao, W., Cassandras, C.G., Belta, C.: Bridging the gap between optimal trajectory planning and safety-critical control with applications to autonomous vehicles. Automatica **129**, 109592 (2021)

87. Xiao, W., Cassandras, C.G., Belta, C.: Learning feasibility constraints for control barrier functions. 2023 European Control Conference (2023)

88. Xiao, W., Hasani, R., Li, X., Rus, D.: Barriernet: a safety-guaranteed layer for neural networks (2021). arXiv:2111.11277

89. Xiao, W., Mehdipour, N., Collin, A., Bin-Nun, A.Y., Frazzoli, E., Tebbens, R.D., Belta, C.: Rule-based evaluation and optimal control for autonomous driving (2021). arXiv:2107.07460

90. Xiao, W., Wang, T.H., Chahine, M., Amini, A., Hasani, R., Rus, D.: Differentiable control barrier functions for vision-based end-to-end autonomous driving (2022). arXiv:2203.02401

91. Xiao, W., Wang, T.H., Hasani, R., Chahine, M., Amini, A., Li, X., Rus, D.: BarrierNet: differentiable control barrier functions for learning of safe robot control. IEEE Transactions on Robotics, IEEE (2023)

92. Xiao, W., Wang, T.H., Hasani, R., Lechner, M., Rus, D.: On the forward invariance of neural ODEs (2022). arXiv:2210.04763

93. Xu, H., Cassandras, C.G., Li, L., Zhang, Y.: Comparison of cooperative driving strategies for CAVs at signal-free intersections. IEEE Trans. Intell. Transp. Syst. **23**(7), 7614–7627 (2022)

94. Xu, H., Xiao, W., Cassandras, C.G., Zhang, Y., Li, L.: A general framework for decentralized safe optimal control of connected and automated vehicles in multi-lane signal-free intersections. IEEE Trans. Intell. Transp. Syst. **23**(10), 17382–17396 (2022)

95. Xu, H., Zhang, K., Zhang, D.: Multi-level traffic control at large four-leg roundabouts. J. Adv. Transp. **50**(6), 988–1007 (2016)

96. Xu, H., Zhang, Y., Li, L., Li, W.: Cooperative driving at unsignalized intersections using tree search. IEEE Trans. Intell. Transp. Syst. **21**(11), 4563–4571 (2019)

97. Xu, K., Cassandras, C.G., Xiao, W.: Decentralized time and energy-optimal control of connected and automated vehicles in a roundabout. In: 2021 IEEE International Intelligent Transportation Systems Conference, pp. 681–686 (2021)

98. Xu, K., Cassandras, C.G., Xiao, W.: Decentralized time and energy-optimal control of connected and automated vehicles in a roundabout with safety and comfort guarantees. IEEE Trans. Intell. Transp, Syst (2022)

99. Xu, K., Xiao, W., Cassandras, C.G.: Feasibility guaranteed traffic merging control using control barrier functions. In: 2022 American Control Conference (ACC), pp. 2309–2314 (2022)

100. Yang, G., Belta, C., Tron, R.: Self-triggered control for safety critical systems using control barrier functions. In: Proceedings of the American Control Conference, pp. 4454–4459 (2019)

101. Yang, X., Li, X., Xue, K.: A new traffic-signal control for modern roundabouts: method and application. IEEE Trans. Intell. Transp. Syst. **5**(4), 282–287 (2004)

102. Zhang, Y., Cassandras, C.G.: A decentralized optimal control framework for connected automated vehicles at urban intersections with dynamic resequencing. In: 2018 IEEE Conference on Decision and Control (CDC), pp. 217–222. IEEE (2018)

103. Zhang, Y., Cassandras, C.G.: Decentralized optimal control of connected automated vehicles at signal-free intersections including comfort-constrained turns and safety guarantees. Automatica **109**, 1–9 (2019)

104. Zhao, L., Malikopoulos, A., Rios-Torres, J.: Optimal control of connected and automated vehicles at roundabouts: an investigation in a mixed-traffic environment. IFAC-PapersOnLine **51**(9), 73–78 (2018)

Index

MIX
Papier aus verantwortungsvollen Quellen
Paper from responsible sources
FSC® C105338

If you have any concerns about our products,
you can contact us on
ProductSafety@springernature.com

In case Publisher is established outside the EU,
the EU authorized representative is:
**Springer Nature Customer Service Center GmbH
Europaplatz 3, 69115 Heidelberg, Germany**

Printed by Libri Plureos GmbH
in Hamburg, Germany